THE EQUATIONS OF LIFE

The **EQUATIONS** of **LIFE**

How Physics Shapes Evolution

CHARLES S. COCKELL

BASIC BOOKS
New York

Basic Books
Hachette Book Group
1290 Avenue of the Americas, New York, NY 10104
www.basicbooks.com

Printed in the United States of America

First Edition: June 2018

Published by Basic Books, an imprint of Perseus Books, LLC, a subsidiary of Hachette Book Group, Inc. The Basic Books name and logo is a trademark of the Hachette Book Group.

The Hachette Speakers Bureau provides a wide range of authors for speaking events. To find out more, go to www.hachettespeakersbureau.com or call (866) 376-6591.

The publisher is not responsible for websites (or their content) that are not owned by the publisher.

Print book interior design by Jeff Williams.

Library of Congress Cataloging-in-Publication Data

Names: Cockell, Charles, author.
Title: The equations of life : how physics shapes evolution / Charles S. Cockell.
Description: First edition. | New York : Basic Books, [2018] | Includes bibliographical references and index.
Identifiers: LCCN 2017056455| ISBN 9781541617599 (hardcover) | ISBN 1541617592 (hardcover) | ISBN 9781541644595 (ebook) | ISBN 154164459X (ebook)
Subjects: LCSH: Evolution (Biology)—Philosophy. | Physics—Philosophy. | Exobiology.
Classification: LCC QH360.5 .C63 2018 | DDC 576.801—dc23
LC record available at https://lccn.loc.gov/2017056455

ISBNs: 978-1-5416-1759-9 (hardcover); 978-1-5416-4459-5 (ebook)

LSC-C

10 9 8 7 6 5 4 3 2 1

CONTENTS

$$P=F/A.$$

Image of Lesser Mole-Rat (*Nannospalax leucodon*)
by Maksim Yakovlev.

PREFACE

SOME OF THE MOST fascinating questions whose answers still remain obscure to science lie at the interface between traditional fields. Of course, disciplines are not real entities. They are artificial constructs made by people. Collecting scientific questions into neat disciplinary boxes is administratively useful but artificial and sometimes intellectually counterproductive. The unguided processes of the cosmos do not recognize these neat divisions, either. There is just the universe, about which a civilization can ask questions.

This book explores one line of thinking that tries to make sense of diverse areas of science that straddle the living and the nonliving, the indefeasible links between physics and evolutionary biology. The connections reflect the reality that life is just a form of reproducing, evolving matter in a universe with many interesting and distinctive types of matter.

The reader should know from the outset that this book is not a sterile attempt to demonstrate that evolution is an utterly predictable product of physics. Historical quirks and chance do play a part, and that point is indisputable. They result in the remarkable plethora of detail and the kaleidoscope of forms that we observe in the great evolutionary experiment occurring on our planet. Travel to the Indonesian islands of Lombok and Bali, and despite their similar size, location, and a mere thirty-five kilometers between them, the fauna of each island is distinct. Life on the islands is the evolutionary progeny of that invisible Wallace Line that carves through the deep waters of the Lombok Strait, placing Bali within the historical trajectory of Southeast Asia's particular evolutionary journey. Bali's forests echo with the calls of Asian woodpeckers and barbets, while Lombok, alive with the shrill cries of cockatoos and honeyeaters, lies within the fold of Australasia's evolutionary umbrella. But lurking within this riot of evolutionary experimentation are unyielding principles of physics. It is those that concern me in this book, principles that have increasingly explained many facets of biology, from the subatomic scale to whole populations—biological observations that were previously assumed to be flukes of history and beyond prediction.

Which features of life are deterministically driven by physical laws and which are mere chance, contingent events decided by a metaphorical role of the dice? This question remains one of the most cogent and interesting puzzles about life and its evolution. I do not intend to provide a definitive answer to this question; I'm not convinced anyone currently has the knowledge to do so. However, I do intend to shed light on the growing understanding of the principles that channel life at all levels of its structure and how this expanding corpus of work shows that life is firmly embedded within the basic laws that shape all types of matter in the universe, much more so than a cursory glance at the menagerie on Earth might suggest.

From this view of life emerge conclusions that some people might find sobering, others might find frightening, and still others thrillingly comforting. For people who share with me a fascination for life on Earth, there is something reassuring about our increasing ability to demonstrate that the apparently fathomless profusion of living things on this planet conforms to simple principles that apply to all other types of matter. For those people who also enjoy speculating about what life on other planets might look like (if it exists at all), a conclusion might be that at all levels of its structural hierarchy, alien life is likely to look strangely similar to the life we know on Earth.

As we let go of many of our ancient traditions of thought that have separated life from inanimate matter, we may find that our fear that this will dangerously consign people and other creatures to what is often viewed as the blandness of physics that determines the fate of most matter in the universe is unfounded. Instead, the unity of evolution and physics brings a new richness to our view of life, an appreciation that within the simplicity of rules that govern and limit the forms of living things there is remarkable beauty.

CHAPTER 1

LIFE'S SILENT COMMANDER

A SHORT WALK THROUGH the Meadows in Edinburgh would leave most people in little doubt that life on Earth is a remarkable anomaly in a universe of bland conformity. Trees of various shades of green rustle in the wind, birds take to the air in gymnastics that left the ancients aghast at the agility of these heavier-than-air flying machines, and along the ground run all manner of animals; the smallest ladybugs land on picnicking tourists, while domestic dogs leap and cavort across the grass.

Compare this spectacle with the velvet black emptiness of space seen with some of our best telescopes. Images of galaxies colliding in astrophysical violence, the light of these long-since-dead encounters collected after traveling billions of years through an empty void almost unimpeded. In this vast, infinite vacuum, punctuated by a collection of stars, planets would materialize. And on one planet, tourists would be swatting flies in a meadow below a castle. What could be more of a contrast between the apparently simple laws that govern the gravitational rotation of a collection of stars—mere balls of fusing gas in the rarefied expanse of space with their attendant planets—and the unpredictable leaps and bounds in something as complex as a pet Labrador, the phenomenon of life?

I once heard a distinguished astrophysicist declare that he was glad to be studying stars, since a star is much easier to understand than an insect.

As we peer into outer space, we can certainly find merit and empathy in this view. Even at the intimidating scale of a massive star, we find simplicity. As the star burns, fusing gaseous elements and releasing energy, so these building blocks successively grow in atomic mass. We begin with hydrogen atoms, the universe's most abundant element, which joins with other hydrogen atoms to form helium. Other helium atoms combine to form carbon and so on, until successive layers are formed—oxygen, neon, magnesium; the atomic mass of the elements grow as the products of each new round of fusion are formed. In the center of the star is iron, which cannot fuse to form any other elements. Iron is the last stage of fusion, and the result of this sequence of relatively simple atomic additions is layer upon layer of heavier and heavier elements from the surface to the core of the star. Elements heavier than iron are formed by other means, such as in the catastrophically energetic explosions of stars that herald the spectacle of a supernova.

Compare the onion-like arrangement of elements in a massive star, over a million kilometers in diameter, to our ladybug (a *ladybird* in many places in the world), just seven millimeters long, sitting on the thumbnail of a sleeping tourist in the Meadows.

The little oval ladybug, a beetle, is just one insect among many species that inhabit the planet. (We do not know how many species there are. About a million are known, and there are likely to be many more still undiscovered.) However, this unassuming creature is full of complexity and comprises eight major parts. Its head is one discrete portion and contains its mouth. The ladybug has antennae and eyes for sensing the world around it, with the antennae considered a separate part. The pronotum, a tough protrusion behind the head, protects it from damage. Behind this are the thorax and abdomen, the body section to which the wings and legs are attached. Finally, this complex machine has elytra, wing cases that protect its vulnerable and delicate flying apparatus.

Yet like the star, each of these features is molded by physical laws. The power of flight conferred by wings means the ladybug must observe the laws that govern aerodynamics, as must all flying creatures. And as for its legs—well, why not wheels? Like all land animals, besides snakes and some lizards that lack limbs altogether, ladybugs evolved legs instead of wheels. There are physical reasons for that, rooted in the relative effectiveness of legs in navigating the irregular terrain of our planet. In protecting their gossamer wings, the elytra must observe rules that pertain to the behavior of tough materials, such as abrasion resistance and flexibility.

In all segments of the ladybug, we can identify the principles of physics. The apparent complexity of the ladybug compared with a star lies merely in the greater number and variety of principles embedded within the insect and which it uses to live its life. Evolution is simply a very good process for assembling different principles, which we can represent as equations, into an organism. Any natural environment usually presents multiple challenges to survival. If a physical process leads to a living thing's development of a characteristic that makes the living entity less likely to be eliminated before it reproduces (reproduction being the measure of evolutionary success), then creatures will evolve over time and can be thought of as containers manifesting diverse physical laws.

The menagerie of life, impressive enough even on a casual walk in the Meadows, grows in variety as you explore its forms through time. Equipped with every scientist's favorite improbable toy, the time machine, we might revisit the Meadows 70 million years ago. Then, we would find forms of life very different from today. Like modern birds, the reptilian pterodactyls had mastered aerodynamics, some with ten-meter wingspans. Feathered dinosaurs and strange insects roamed the land, and in ponds or lakes, reptiles, slender and long, achieved mastery in their watery habitats. If we hop back into our machine and return to about 400 to 350 million years ago, we would find Edinburgh in the site of immense volcanic activity, thick mat-like conglomerations of microbes growing here and there. Across the land, between the volcanic cones, the earliest land plants, the *Cooksonia*, stake their claim to the new habitat. Scurrying among their short knobbly stems of just a few centimeters' height were early insects and the now extinct eight-legged beetle-like trigonotarbids. Return just a few tens of millions of years later and you would have seen four-legged *Pederpes*, a forerunner to modern land vertebrates, awkwardly shifting its one-meter-long slithery body through the undergrowth, peering this way and that with its triangular head to chart its way through.

But impressed though we may be with all the wonderful life forms we have observed in our excursion through time, there is a strange familiarity about all these living things. Their shapes and forms, although different, share fundamentally the same types of solutions to living as we see in modern forms. These similarities are not merely an artifact of evolutionary descent. The growth of early plants against gravity, the size of bones that hold up a dinosaur, the sleek shapes of water-bound animals, and the features of wings that allow a pterodactyl to fly lead to evolutionary forms similar to modern-day organisms faced with the same unyielding laws.

The complexity and sheer diversity of life in time and space could convince anyone that life represents something quintessentially different from physical processes, a divergence of form that transcends the simple principles that seem to fashion the apparently predictable structure of the inanimate world.

Yet physical laws restrictively drive life toward particular solutions at all levels of its assembly. The outcomes are not always predictable, but they are limited. It does not matter at what level those requirements are operating, from the subatomic to the population scale: the results are various, but not boundless.

Even at the smallest scale, we can see these narrow channels of evolution. Consider molecules, such as the proteins, from which the eclectic mixture of monsters on Earth is assembled. Like life at the larger scale, proteins do not display untrammeled potential. Rather, in observing the limited ways that proteins (including enzymes, biological catalysts) can be folded together, some scientists have argued that these configurations reflect a set of forms analogous to the perfect and unchanging forms of things espoused by Plato. To some people, such a view seems to contradict the Darwinian view of life, which emphasizes natural selection's tendency to produce apparently unlimited variety.

Science can sometimes be unnecessarily polarized, and, of course, challenging the Darwinian view is always popular, edgy, and controversial. In the synthesis I present here, there is no challenge to Darwin's basic precept that natural selection can fashion an extraordinary range of creatures or even proteins. I merely illustrate how this process is limited in the basic pattern of its products, not just at the level of the organism, but in any part of its construction, from populations down to proteins and down to the atomic level. I gather evidence from what has become an impressive corpus of work by many researchers to show how physical principles sharply narrow the scope of the evolutionary process at all levels of life's structure.

My view is underpinned by a simple proposition. Evolution is the process by which the environment acts as a filter to select units of organic mass in which a mosaic of interacting physical laws are optimized sufficiently to allow for reproductive success. The environment in this context can include a whole range of challenges—from weather phenomena such as storms to the appetites of predators—that might prevent an organism from reproducing. Evolution is just a tremendous and exciting interplay of physical principles encoded in genetic material. The limited number of these principles, expressed in equations, means that the finale of this process is also restrained and universal.

Equations are merely a means to express in mathematical notation the physical processes that describe certain aspects of the universe, including features of living things. The phrase *equations of life* is shorthand for this growing capacity to use physical processes, and often their mathematical formulations, to describe life at different levels of its hierarchy. Throughout this book, I will provide some examples of these equations, but I do not expect you, the reader, to have to understand their nuances and details or how to use them. I show them to illustrate how physical principles that underlie evolution can often be expressed in these conveniently shortened mathematical forms.

That the laws of physics should bound life is hardly a controversial statement. The ladybug in the Meadows does not exist outside the same principles that govern the formation of the Sun that warms it on a sunny day. Life is very much part of the universe: it cannot operate outside its rules. Yet although this observation seems trite, we are often remarkably unwilling to accept the extent to which life is tightly constrained by physical laws. When observing living things, we can easily forget that the rules that govern their architecture set a harsh perimeter; the extravagance of life can appear, to the unreflective observer, to be limitless in its variety.

For me, what is surprising about the journey through the different scales of life's structure, accompanied by our growing knowledge, is that life is much more amenable to description in terms of physical processes, and thus simple mathematical relationships, than once was thought, and that these principles are now being elaborated at many different levels of its hierarchy. These insights also suggest that life is much more narrowly circumscribed than those who favor the role of chance and history would like to think. Accordingly, life is more predictable and potentially universal in its structure than is sometimes assumed.

In their details, living things do show apparently illimitable embellishments. The vastness in the details of living things has probably caused a divergence between the biological and physical view of the universe. Yet if we return to our facile observation that biology operates within the laws of physics, then we should be able to more comprehensively reconcile this division.

When I joined a university physics department a few years ago as an astrobiologist, I was asked to teach an undergraduate physics course called Properties of Matter. For me, with a background in biochemistry and biology, a semester of this material would be unpalatable without some biology, so I set about modifying my task by using biological examples to illustrate the physical laws and ideas I needed to teach. The inclusion of some biology improved my own motivation, and I also thought that doing so would be interesting for undergraduates.

It was not a difficult task to find these examples. At the molecular level, the van der Waals forces that hold molecules together—these feeble forces from the inherent polarity in molecules make the molecules behave like little bar magnets (even the unreactive noble gases such as neon can behave like this)—can be illustrated with a gecko. These agile desert lizards have an abundance of tiny hairs on their toes; the hairs allow the combined van der Waals forces on all four feet to hold the creature fast on a vertical surface, allowing it to run up a shiny glass window with ease.

The two strands of the genetic material DNA, the molecule that encodes the information in your cells and all other cellular life, are held together to make the familiar double helix by links called *hydrogen bonds*. The forces involved in these links are just enough to hold the strands together and maintain the integrity of the molecule, but just weak enough that the two strands can be easily unzipped when the cell is dividing in two and the information in DNA must be copied. The replication of DNA and the architecture of its multiplication can be understood as the forces between atoms.

At higher levels of its hierarchy, biology still came to the fore. In explaining phase diagrams (graphs that show the state that matter adopts at given pressures, temperatures, and volumes), I found some illustrations from the world of biology useful. The fish that swim unmolested by predators and in the comparatively warm water trapped beneath the ice on a frozen wintry pond take advantage of the negative gradient of water's melting curve on a phase diagram. Put simply, when water is frozen, it becomes less dense and floats. Fish that remain active in the winter have evolved to cope with living in the habitat under ice—their behavioral evolution is constrained by some simple facts about the behavior of water that can be manifested in a phase diagram.

Even at the macroscopic scale, physics both explains biological systems and constrains their operation. When clarifying how large creatures travel through water, we are confronted with questions such as why fish lack propellers—what physical laws make a flexing body a better way to get through the ocean and away from a shark than a propeller, the solution of choice to human engineers? The behavior of fluids and the objects that travel through them provides extraordinarily tight constraints on the organisms that can evolve and the solutions they find to live within these constraints.

After teaching this course, I was surprised not by how we could find biological examples of physical laws in action, but by how deeply simple physical rules fashion and select features in life at all levels of its hierarchy, from an electron to an elephant. I was well aware of how physics can shape whole organisms, but I was awed by the sheer pervasiveness of the reach of physical principles, like tentacles, stretched through the entire fabric of life. And despite the inherent uncertainty swirling around subatomic particles in the quantum world—uncertainty that might reasonably make a cautious physicist wonder about how confidently we can bring biology and physics together—the shape and chemical composition of Schrödinger's cat and the height of Werner Karl Heisenberg himself are highly predictable, convergent features of physical principles operating in biology.

Sometimes, scientists use the oceans as an analogy to evolution. Different animals represent islands of biological possibility, where solutions to successful adaptation to the environment are constrained by what is physically possible and what an organism already has on hand: its history. Between these islands, there are vast oceans of impossible solutions that life must navigate between to find new islands of possibility. Its seems extraordinary that life manages to home in on these islands and that living things seem to arrive at the same haven, like a party of separated shipwrecked seafarers that find themselves marooned on the same deserted outcrop in the middle of the Pacific Ocean. How is it that two animals, such as a bat and a bird, home in on the same functional solution to flight? This convergence cannot be easily explained by a common ancestor, since their ancestors lacked wings, a fact borne out by the very different wing anatomy in the two creatures. However, there is nothing uncanny about life's ability to land on the same solutions. Impossible solutions are impossible solutions, which means that the ocean of impossibility does not exist at all.

We might instead try to visualize the physical aspects of evolution as like a chessboard. Each square is a different environment, a different set of physical conditions to which life must adapt. When a living thing moves across the board, it automatically finds itself in another space to which it must adapt using a range of well-defined physical laws. For example, the laws of hydrodynamics that enforce certain forms in a fish would be replaced by the dominance of new rules when it crawls onto land. Limbs that allow movement against a more dominant influence of gravity and equations that determine the rate of evaporation as the midday sun mercilessly tries to desiccate our new denizen of the landmasses become some of the shapers of evolution. But there is no intermediate ocean of impossibilities. There are only physical principles seamlessly operating together in different combinations and magnitudes in different environments. Life moves from one environmental condition to another, those laws operating all the time to select successful conformists to physics, while the environment or competitors ruthlessly eliminate the forms whose adaptation to the unwavering requirements of these laws fails to allow them to reproduce.

There is a distinction worth making here. The ocean analogy works rather better when we think about how effectively creatures are adapted to their environment. In the extreme example, an insect born with a missing wing is likely to be severely handicapped in its ability to succeed in the evolutionary game. The idea of organisms occupying a vast landscape where islands and the peaks of mountains represent organisms best adapted to their environment and the plains and oceans between as the organisms less well adapted and less likely to succeed forms the basis of the concept of adaptive landscapes. However, there is nothing strange about life's ability to find similar evolutionary solutions to environmental challenges. There is no empty space to explore. Living things just move from one place to another; when the physical laws confront them, they must adapt to reproduce. If they do not, we never see them again. Those physical laws often demand similar solutions.

In this book, I do not expect the reader to be surprised that biology and physics are inseparable, that physics is life's silent commander.

Instead, I intend to illustrate the wonderful simplicity of life from population to the atomic scale. I also suggest that these laws are so ingrained, from the atomic structure of life to the social behavior of ants, that life elsewhere across the universe, if it exists, will show similar characteristics.

Surely, though, we might say, "Life cannot just be about physical principles. What about the cheetah that chases the gazelle? Not merely a physical effect on the gazelle, but a true biological interaction." The cheetah that races across the African savanna to catch the hapless gazelle for its next meal is exerting a selection pressure on the gazelle, and this pressure is, at the level of the biological response, physical. The gazelle will survive this encounter if it can outpace the cheetah. Whether the gazelle can escape depends on how quickly it can release energy in its muscles or how deftly it can twist and turn as it seeks to evade the oncoming predator. This ability is itself a product, among other things, of the forces that the knees of the gazelle can endure and the torsion that its leg bones and muscles can accept as it seeks freedom. These factors ultimately are determined by the structure of muscles, bones, the acuity of eyesight, and so on. Either the gazelle will survive to get closer to reproductive age, or it will not. This selection pressure cares not that the cheetah is another biological entity. It could just as well be a fast-running robot built in a physics laboratory at the University of Edinburgh programmed to run across the savannas of Africa, randomly intersecting and killing gazelles. The only matter of importance is whether the biological, and ultimately physical, capabilities of the gazelle allow it to survive the cheetah and what adaptations in muscular properties, bone strength, and other factors allow its offspring to be the successful successors.

The points I make above apply equally to the evolutionary changes that occur in organisms not just from selection restrictions in the environment, such as predation, but also by new expansive opportunities provided, for example, by unexplored habitats and food resources. Many of these changes, both in the short term and ultimately in evolutionary terms, projected onto living things in their environment may be caused by fellow biological travelers on Earth. However, the adaptations required to ultimately survive or exploit the changes in the environment or other organisms are often tightly channeled by physical principles.

All these adaptations are, of course, bounded by the restrictions that may be imposed by the prior shape and form of the organism's ancestors or in its developmental patterns. These historical architectures and limits in how living things can develop and grow are in themselves boundaries set up by previous evolutionary selection, and these boundaries merely constrain how an organism can respond to the full set of physical laws theoretically available to it and imposed on it. They narrow the field of play further.

There is a question that might be lurking in the mind of reader. You might be wondering, "But what *is* life?" After all, the preceding discussion has rather assumed we agree on what life is. The question of what defines life has occupied the minds of many good people for a long time. But for the purposes of this book, I do not need to advance that discussion. For simplicity's sake, I take as implicit in this book a convenient working definition of life, which is essentially that living matter is material capable of reproducing and evolving, consistent with a definition made by biochemist Gerald Joyce, that life is a "self-sustaining chemical system capable of undergoing Darwinian evolution." The capacity to evolve, that is, Darwinian evolution, is the feature of life that allows organisms to change over time and become better adapted to their environment. On a more easily understood level, on the Earth, this capacity includes almost all the familiar life forms, including the eukaryotes, the domain of life that embraces animals, plants, and many other organisms such as fungi and algae, and the prokaryotes, within which the bacteria and archaea (another branch of single-celled organisms) reside.

We could argue that the word *life* is merely a human categorization, something that will never yield to concrete definition. Life might just be an interesting subset of organic chemistry; it is a branch of chemistry that broadly deals with lumps of carbon compounds that happen to behave in complex ways. Its capacity for reproduction leads to evolution as environmental forces act on this reproducing material. Life's apparent persistence on the planet is a product of the evolution of a genetic code within the reproducing material; this code allows for modification and variation in many reproduced units of that matter. Selection pressures act in different environments to whittle the variants down to the successful ones that are subsequently reproduced and distributed into new conditions.

However, whatever we decide about life, whatever the definition or concept we choose, any of these possibilities is entirely consistent with simple physical laws. In his engaging 1944 book, *What Is Life?*, Nobel Prize–winning Austrian physicist Erwin Schrödinger famously described life in physical terms as possessing the attribute of extracting "negative entropy" from the environment, a slightly unfortunate phrase as it has little formal meaning in physics. However, it was a phrase he chose to capture the idea that life seems as if it is working against entropy, which is the tendency for energy and matter to be dispersed and dissipated into thermodynamic equilibrium. Entropy is a basic attribute of matter and energy encapsulated by the second law of thermodynamics, which recognizes this tendency for things to achieve such an equilibrium. In many cases, this attribute equates to things becoming more disordered. In Schrödinger's view, life was in a struggle to fight entropy.

Life tends to create order in a universe ultimately prone to disorder. This attribute perplexed Schrödinger and has seemed mysterious to generations of thinkers. When a lion cub grows and eventually reproduces, all the new matter bound up in that adult lion and its offspring represents more ordered, less randomly dissipated energy than when the lion was a small cub nipping at its mother's heels. Indeed, for a long time, it was something of a challenge to biologists and physicists to explain why life seemed to be doing something apparently in violation of the laws of physics. However, when we look at life in another way, rather than viewing it as something anomalous and almost fighting the laws of physics, we can instead see it is a process that accelerates disorder in the universe—very much in line with physical processes that describe the cosmos. The best way to explain this idea is to use my lunch sandwiches.

If I place my sandwiches on a table, provided they are left alone, it will take a very long time for the energy in their molecules to be released. Indeed, the energy in the sandwich may not be released until it ends up in the Earth's crust from the movements of the continents during plate tectonics, the sandwich crashing down into the depths of the Earth, heated to great temperatures in the far future, when its sugars and fats will be broken down into carbon dioxide gas. However, if I eat the sandwiches, within about an hour or two, their contained energy will be released as heat in my body and some carbon dioxide exhaled in my breath, with some portion of it being used to build new molecules. In essence, I have accelerated, greatly, the dissipation of the sandwich into energy. I have enhanced the rate at which the second law of thermodynamics, which drives the universe toward disorder, has had its way with the sandwiches. Of course, if I leave my sandwiches on my table, they will go moldy and be eaten by bacteria and fungi that land on them—these organisms will have merely beaten me to it in dissipating the energy of the sandwich into the rest of the universe. Mathematical models show that this idea is not mere whimsy, but that the process of life and its tendency to grow, expand in population, and even adapt can be described by thermodynamic rules.

Living things show extraordinary local complexity and organization, but the process they are engaged in is accelerating the dissipation of energy and the rundown of the universe. Local complexity in organisms is an inevitable requirement to construct the biological machines necessary for this dissipative effect to occur. As the physical universe favors processes that more rapidly dissipate energy, then life is contributing to the processes resulting from the second law, not fighting it. At least, that is one way to view the phenomenon of life. Seen from this perspective, it is easier to understand why life is successful.

Ultimately, of course, when there is no more energy to dissipate or when environmental conditions become unsuitable for life as the oceans all boil away in the searing sky of a more luminous Sun a couple billion years from now, these local oases of complexity that once seemed to defy the second law will do so no more. They too will be destroyed.

This apparent detour relates to us simply because it underpins the idea that life is very much a physical process at work. Living things are collections of molecules behaving in a way that is consistent with, and encouraged by, the laws of physics. We would expect it to be elsewhere across the universe. Within this overarching behavior, the living things carrying out this process are themselves subject to the laws of physics. In these pages, I am less concerned with prolix and otiose deliberations on the definition of life and more interested in the universality of reproducing and evolving matter that we choose to call life.

The more we learn in physics, chemistry, and biology, the more we are confronted by the simplicity of rules that govern the universe and their unexceptional character. It has been something of a theme through the history of science that major paradigms have overturned the exceptionalist view of our place in the universe. The Earth as just one planet circling the Sun and the descent of humans from apes are two of the most traumatic conceptual changes to our worldview in the last few hundred years. These ideas replaced the geocentric view of the universe—the Earth at the center of our Solar System, populated by people very special and separate from the rest of the animals.

That biology conforms to physical laws raises fundamental questions about the wider universal view of biology: If life exists elsewhere in the universe, will it be like life on Earth? Is the structure and form of life unexceptional as well? At what level of organization could life elsewhere be the same? Is the choice of elements in the ladybug's leg the same in another galaxy? What about the molecules the atoms come together to form—would the molecules that build and shape the ladybug leg be the same? And what about the ladybug itself? Are there other ladybug-like creatures in another galaxy? Could it be that ladybugs, at all levels of architecture, are unique to Earth?

If physics and biology are tightly coupled, then life outside Earth, if such life exists, might be remarkably similar to life on Earth, and terrestrial life might be less an idiosyncrasy of one experiment in evolution, and more a template for much of life in the universe, if it exists elsewhere. Such an assertion would imply predictability, the hallmark of a good scientific theory.

A favored trope among science-fiction writers is to imagine any number of extraordinary life forms inhabiting other planets and to contend, therefore, that we are limited in our imaginations and, consequently, that we cannot make sensible predictions.

As early as 1894, in a *Saturday Review* article about alien life, science-fiction writer H. G. Wells reflected on earlier suggestions that silicates (the silicon-containing materials that make up rocks and minerals) might do interesting chemistry at high temperatures: "One is startled towards fantastic imaginings by such a suggestion: visions of silicon-aluminium organisms—why not silicon-aluminium men at once?—wandering through an atmosphere of gaseous sulphur, let us say, by the shores of a sea of liquid iron some thousand degrees or so above the temperature of a blast furnace."

He is not alone. In 1986, Roy Gallant wrote *Atlas of Our Universe*, a well-known exposition of the possibility of the limitless potentialities of life, for the National Geographic Society. The book contains a wonderful plenitude of imagined life forms in our Solar System. The Oucher-pouchers are large bags of gas that prance around on the surface of Venus and cry "ouch" every time they hit the surface, baking at 460°C. Their counterparts on Mars are the Water-Seekers, long, slender creatures like extended ostriches that sport vast furry ears with which they can enclose themselves in the cold Martian nights and winters. A giant carapace over their heads protects them from ultraviolet (UV) radiation, and with their long proboscises, they dig deep into the Martian subsurface to find water. The imagination reaches far beyond these two worlds. The Plutonian Zistles are intelligent ice cubes on Pluto (the National Aeronautics and Space Administration [NASA] New Horizon mission, perhaps glinting briefly overhead, presumably changed their culture for good), and the Stovebellies of Saturn's moon Titan combust material inside themselves to maintain warmth at a chilly −183°C. They propel themselves through Titan's hydrocarbon-rich atmosphere by the unedifying means of emitting bursts of gases from their rear ends.

None of Gallant's creatures have ever been observed, and that is an interesting fact. Assuming (and this is a big assumption) that life would emerge on other planets if the conditions were right, these novel biochemistries and creatures are, not insignificantly, absent in our Solar System—life forms that would have merely adapted to the different conditions found on those worlds. On most of these worlds, the conditions are so extreme that, according to our knowledge of the limits to life on Earth, we would predict that none of these planets and moons could today have complex multicellular life on their surfaces. That is what we observe. What we see on Venus, for instance, fits our predictions based on our knowledge of the boundaries of growth of terrestrial life—boundaries established by physical laws.

We do not yet have another example of life with which to test whether our biosphere is exceptional. Consequently, many observers might say that the question of whether life on Earth represents a universal norm can be nothing more than speculation, unbounded conjecture of the type that makes interesting conversation at the coffee table, but little more. However, this observation is inaccurate. The principles provided to us by physicists reveal the foundations of what is possible. Observations of the universe from astrophysicists can tell us about the preponderance of elements such as carbon and of molecules such as water; this information can yield insights into how common the chemical building blocks of life may be throughout the cosmos. From chemistry laboratories, our extensive knowledge about the reactive potential of elements in the periodic table and their ability to form complex structures can tell us about how universal the chemistry of life might be.

The biophysicists have much to tell us about molecules that have evolved independently in many organisms on Earth and allow us to question how universal the rules for doing chemical reactions in cells might be. The microbiologists' knowledge of life in extremes informs us about the physical boundaries of life and whether these are likely to be universal. From the paleontologists, we are given a vista across the life forms of the past. How similar or different are they from those alive today, and what might explain these observations? Planetary scientists collecting information about other worlds tell us whether, with their cameras and other instruments, they find conditions potentially supportive of life. We can compare the biological status of these worlds with our expectations.

From these disciplines, we can gather an abundance of information to build a hypothesis about the nature of life. In this book, in investigating the link between physics and life, I also explore the idea that life is universal at all levels of its hierarchy. By this, I do not necessarily mean identical. Ladybugs may not be the same on other worlds as on Earth, but the solutions used by living systems to proliferate on the surface of a planet might be broadly similar, from the way they use a subatomic particle, the electron, to gather energy right through to the behavior of their populations. We will know life if we eventually find it, and it will be recognizable as very akin to life on Earth.

It is apposite to give Charles Darwin the last word of the first chapter. In the finale of his *Origin of Species*, he summed up his feelings by declaring that he could see a certain grandeur in the evolutionary view of life. We might also remark that there is a beautiful simplicity. As physical laws, unyielding and unswerving, work their way through every form of life, extinct and extant, there is a breathtaking similarity in the products of evolution, a resemblance molded by the very laws that have shaped our universe for over thirteen billion years.

CHAPTER 2

ORGANIZING THE MULTITUDES

WHEN I WAS A boy of about eight, I was a typical daydreaming child. I would sit on the Victorian stonework, leaning against the black iron railings, and with my small magnifying glass, I would focus the rays of the sun on an oblivious ant going about its tasks. With my death ray, I would chase the little creature across the irregular pitted surface until I caught it in the glare and it spat and fizzled.

Ant chasing was an extracurricular activity at a typical English boarding school, and it was preferable to some of the others, including learning more Latin. I dare say it remains a macabre pastime for inquisitive, slightly destructive children to this day. In my juvenile unpleasantness to these innocent insects, I was party to their miniature world. I saw on many occasions the long, regular lines of ants tramping back and forth across the stones, some moving slowly, others fast, some with pieces of food, and a few with the carcasses of their fallen comrades. Every now and then, two of the scurrying forms would clamp heads as if exchanging instructions and then part, running off with hurried intent in opposite directions. What were they saying? The social activities of these ants fascinated me, and more often than not, I would prefer to just sit diligently and watch them.

But I saw something else. In my preadolescent activity, I witnessed the delicate nature of life. By merely taking the natural light from the sun and magnifying it just a few times, I could transform a living, intricate machine of organic matter into a blazing inferno. Life really was tenuous, poised at the boundary of physical extremes that, with a mere change in their magnitude, could define the difference between life and death. These creatures, like all of us, lived in a world at the mercy of hard physical limits.

Nevertheless, within these constraints, the ants went about their business. Watching them coalesce, exchange information, and organize could convince anyone that what was at work here was nothing short of social organization. A vast society of insects, merely on a scale smaller than us, worked toward their goal of constructing their nest and ensuring that they had enough food to perpetuate their colony. For many years, this top-down society was how it seemed to scientists. The queen ant, safely ensconced in a chamber within the nest, was further proof that this incredible collective effort was under the control of a monarch, a figurehead that directed and controlled the many instructions that must be needed to coordinate hundreds, thousands, and sometimes millions of ants toward a single, unambiguous task.

It is easy to see how this phenomenon led many to question how such a tiny thing, the queen ant, even at her often-bloated size, could possibly contain, let alone process the astonishing amount of information needed to operate the ant society. Ant civilization attracted the attentions of many biologists and animal behaviorists, such as American scientist E. O. Wilson, whose work from the 1970s on insect societies helped found the field of sociobiology.

A fascination with insect societies and the draw to understand what managed their multitudes caused a new group of scientists to become engaged with ant organization. Physicists, who are wont to avoid the dizzying complexities of things so unconstrained as ants, took an interest in the creatures. A collaboration emerged between biologists and physicists. They asked some different questions: Are these societies really so complex? Are they fashioned by flows of information and instructions beyond the realms of our computers? Are they under the whims and dictates of their queen, a leader we may never fully fathom?

What they found was remarkable.

Ant nests are complicated structures whose extent and detail can reach colossal proportions. A metropolis of nests containing an estimated three hundred million worker ants and a million queens was found on the coast of Hokkaidō, Japan, in 2000. Not a single nest, this labyrinthine construction of forty-five thousand nests was connected by shafts and tunnels and covered an area of over 2½ square kilometers. If such a city were to be built by humans on our own scale, we would require numerous architects musing, deliberating, and planning with someone who could oversee the whole project and could be relied on to keep the entire enterprise on track.

Yet among the ants, it seems, such vast empires can be constructed using the simplest rules.

Deep underground, an ant carefully and gently removes one soil grain at a time, dragging it away and dumping it to one side. Quietly and with seeming intent, it starts on a job too big for a single ant, so it releases some chemicals, pheromones, that attract a neighbor to help. Now two ants are busy working away, removing grains and starting the task of building a new chamber. They too need help, so they recruit two more ants, and those four ants recruit four more. Now there are eight. Quickly, we have what is known as a *positive feedback effect*, a near exponential growth in the number of ants now steadfastly dragging away grains of soil. At last, we have some significant effort and the chamber begins to grow at a sensible rate. Over minutes and hours, the new home takes shape.

But there is a problem: there are not an infinite number of ants available. Other chambers are being built, adding to the pressure for workers across the expanding empire. As the chamber grows, the ants working on it become more dispersed across its surface. The recruitment of ants slows down, and as it does so, a negative feedback sets in. Fewer ants mean reduced pheromone emissions and therefore still fewer ants. Building of the new chamber grinds to a halt. But no worry, because next door, another ant has started a new hole next to a tunnel full of ants. And so the process repeats: in little holes across the nest, brand-new chambers are formed. Now with all this fresh space available, the nest can accommodate more ants, so as the volume of the nest grows, the total population of the colony will also swell, keeping pace with the volume of the nest.

Take these simple ideas of positive and negative feedback effects between individual ants meeting and greeting each other in their fossorial wilderness, and write them into a computer program. Now you can recreate the chamber-building activity of ants and even predict the growth of the whole colony.

Remarkably, no architect is needed for this task, no grand designer to draw the ant nest on a board and to direct and supervise the workers on the job. Despite the overpowering wish to draw some sort of parallel between the impressive scale and collective effort these insects master and the building of the Egyptian pyramids, there could not be a bigger difference. The ant nest can be predicted with nothing more than simple rules operating between individual ants. The queen provides the focus of the nest, the source of eggs and new workers, but the everyday tasks of building the nest are the product of basic interactions between lots of busy ants.

The consequence of this order is that some of these antics can be written down in relatively straightforward equations. Often in the natural world, in physical, chemical, and biological systems, a *power law* explains the relationship between things. Put simply, it means that one item we might be measuring, such as the volume of an ant nest, changes in proportion (as a fixed power) to something else, perhaps the number of ants, with the simplest expression being:

$$y = kx^n$$

where x is one thing we might measure (say, the volume of the nest), y is what we want to know (say, the number of ants), and n is the number (the power, hence, a power law) that scales the relationship between them (which can be measured). For example, for the ant species *Messor sancta*, the value of n is 0.752. The value k is another proportionality constant that can be worked out for any given process.

Power laws come about because of some inherent link between two things that are being measured, and often that link is rooted in a physical principle. For our ant example, the more ants there are, the more grains of sand or soil they can move. Since the collected three-dimensional grains essentially amount to the total volume of the chambers in the nest, it perhaps is not surprising that all other things being equal, the number of ants is related to the volume of nest they build.

Not confined to ants, power laws scatter through biology from the largest to the smallest scale. The laws appear in other places as well, since their ubiquity underscores the regularity in life. In quite different places, we find the same mathematical relationships. The laws of the ants are written in the same formula as other features of living things.

Perhaps best known among power laws is Kleiber's law, named for Max Kleiber, a Swiss-born physiologist. He measured the activity of a variety animals and found a simple relationship between the metabolic rate, essentially the energy the creature is burning, and its mass:

$$\text{Metabolic rate} = 70 \times \text{Mass}^{0.75}$$

This equation tells us that large animals have greater metabolic needs than do smaller ones. A cat has a metabolic rate about thirty times that of a mouse. This relationship makes sense, since a large animal has more mass to keep going. However, the power law also tells us that smaller animals have a higher metabolic rate for each part of their volume than do larger animals. Smaller creatures tend to have a higher proportion of "structure" such as muscles than fatty reserves than larger animals have. They also have a high surface area relative to their volume and so they will tend to lose heat more easily, burning up more calories per unit of weight than larger animals burn.

The exact physical underpinnings of Kleiber's law and many other so-called allometric power laws that relate the size, physiology, and even behavior of living things are becoming better understood.

Their elevated status to "laws" would make many physicists wince. Most of these mathematical observations do not express some fundamental law like Newton's laws of motion; rather, they are general relationships. However, these intimate links, like many other power laws in biology, speak to us of the underlying order in the biological world, the interconnectedness, from populations of ants to the size and physiology of living things, that ultimately must conform to real physical laws. Many fixed relationships between the features of living things such as metabolic rate, life span, and size of animals that conform to power laws can be explained by the network-like properties of life.

Within the phenomenon of the ant chamber, there is a beautiful example of how complexity can emerge from populations of organisms with simple rules. Put many ants together that interact, and the to-and-fro of their exchanges will lead to patterns. At their core, the interactions are elementary, but mixed and matched, they lead to variegated behaviors.

Attempting to reduce the tangled complexity of populations of organisms, from ants to birds, to more tractable physical laws has fallen under the realm of a part of physics sometimes called *active matter*. This field strives to fathom how matter behaves when it is far from equilibrium, when it has not settled down into a stable and sometimes inactive state. For most of us, being "out of equilibrium" is synonymous with disorder and imbalance. Yet, physicists have found that when systems are far from being in a settled state, rather than disorder, ordered patterns sometimes emerge, and this order can drive biological processes.

In a landmark paper published in 1995, one early attempt to ignite the study of active matter, Tamás Vicsek at Eötvös Loránd University in Hungary developed a straightforward model of hypothetical particles bouncing around and occasionally meeting each other. He found that at low density, these virtual creatures, or blips of data, behaved randomly. Their concentration was just too small for anything noteworthy to happen. However, bring them together at high enough density, and now they move in a way that is influenced by the movements of their neighbors. Mutual interactions cause collective patterns and behavior to emerge. A shift, a phase transition, from one state to another dramatically occurs. These early beginnings in the field of active matter showed how grand things can happen from simple designs. A growing interest in self-organization in living and nonliving systems followed.

Biology is no doubt a special part of active matter. Living things have history, evolutionary quirks, behavioral specialisms even, that make them not mere particles bouncing off each other like atoms of a gas in a box, but more complex and, to some extent, unpredictable entities. Despite these idiosyncrasies, many features of the biological world at the scale of whole populations are successfully reduced to principles that are more transparent. From the swarming of bacteria to the flocking of birds, equations can be derived that help predict behaviors seen in the natural world. Vicsek's elegant paper hinted at the physical underpinnings of collections of entities in evolution's great experiment.

The feedback loops that direct the growing ant empire also decide how those same ants will get a meal. Outside our ant nest, a delicious juicy orange has just dropped from the branches of a tree. Within days, under the warmth of the sun, it begins to rot, oozing its sugary interior onto its surroundings. An ant, scouting around outside the nest, its antennae feverishly flicking, picks up the scent of the moribund fruit. It stumbles across this treasure trove and swings into action. Scurrying around the orange, it bumps into one of its fellow workers and, in a brief meeting of heads, instructs the other to return to the nest and recruit more workers. Soon, a trail is established to the nest, ants darting back and forth along the trail, great globs of sugary fluid in their mandibles. As each ant in the trail recruits others nearby, the numbers multiply rapidly and soon we have a miniature road crammed with them dashing back and forth.

As the ants smother the orange, no amount of extra workforce will be much use. Now there are too many cooks in the kitchen. Soon the orange, dismembered in the feasting mandibles of the colony, runs dry and thus the numbers of ants recruited to the orange declines. Other ants, ensuring that the nest does not rely on only one orange, react to "keep-away" messages. These black sheep of the colony, if you will, deliberately head off in new directions to find new food sources. Eventually, the orange is depleted and the trail dies away. In the appearance and disappearance of ant trails, we have no queen ant sitting in her chamber with a map, planning new excursions to find food, drawing lines on a grid, and instructing her minions to systematically scour each square for food. Instead, plain rules, beginning with a lone scout ant trekking across the home turf, lead to mathematical processes that end in food.

Like other aspects of the ant world, we can even write this entire scenario in an equation:

$$p_1 = (x_1 + k)^\beta / [(x_1 + k)^\beta + (x_2 + k)^\beta]$$

where p_1 is the probability that an ant will choose a particular trail to run down. The probability is predicted using x_1, which is the amount of attracting pheromone on that trail, which may just equate to the number of ants already on the trail. The variable x_2 is the amount of pheromone on an unmarked trail, which an ant might follow instead. The variable k is the attraction level of a pheromone on the unmarked trail, and β is a factor that takes into account the nonlinear behavior of ants, in essence some of their social complexities and behavior that vary from species to species. The higher the value of β, the greater the probability that an ant will go down a trail even if the trail has only slightly more pheromone.

This is the equation of feeding ants. Here, in essence, we have an equation that predicts where ants like to go for food.

A human analogy to this whole episode is the arrival of a new artisan cheesecake shop in Edinburgh. Delicious new cheesecake, handmade at that, is a delight for city dwellers to offer at their summer luncheons. Delia accidently stumbles across the shop and buys some for her next gathering. Her guests are delighted, so she tells her friend Sophia. Now Delia and Sophia are both telling their friends, and soon, everyone is calling everyone else. There is a run on the cheesecake shop. It's the place to go. A cheesecake feeding frenzy engulfs Edinburgh. Soon, however, there is no one left in Edinburgh to call. Everyone knows about the cheesecake shop. The number of people dropping in at Bruntsfield Cheesecakes, begins to plateau. But there is worse. Now cheesecake is no longer chic. Soufflé is the order of the day, and a new shop opened up on George Street that does some pretty nice stuff. Those in the know now call their friends to get ahead of the game. Avoiding the cheesecake shop in favor of the new trend, the shop's clientele declines. As the shop cuts back on making cheesecake, this sets in motion an even smaller demand and the cheesecake shop is all but abandoned.

Delia and her cheesecake or soufflé preferences look like a complex social arrangement, but they follow simple rules. She and her friends have received no instructions from Edinburgh Council (or the queen herself) on whether to buy cheesecake or soufflé. In the ant world, without the real complication of the somewhat intricate social mores of humans, these simple feedback processes also drive the ants to switch from one food source to another.

The world is never as straightforward as a single orange. Perhaps several oranges have fallen from that tree. Faced with a tantalizing choice, even the smallest fluctuations in the number of ants running around could lead to one orange or another being chosen first. So predictability comes from the equations—we can define the rules that decide in principle which trail an ant will go down, but there is unpredictability in how the equation is manifested and the exact trail it will work its effects on. In the complex variations of the natural world, these small fluctuations play an immensely important role in shaping behavior, and no doubt they contribute to our sense that living things are inherently unpredictable, different from inanimate objects.

Other occasions cause the rules to be less easily discerned. A particularly large ant colony may have so many individuals that they simultaneously swarm many oranges, tearing them apart in a feeding frenzy. Under these conditions, our delicate feedback effects are all but gone to the wall. And, of course, the environment itself will mess up those nice, elegant equations. Put one of these oranges in a crack in the ground or under some particularly cumbersome vegetation, and the trails and feedback processes suddenly become motley and tortuous. Nevertheless, beneath these quirks, the equations of ants work their way.

The feedback systems operating in the nest might even help explain another enchanting and mysterious feature of animals: synchronicity. This quality appears not merely in ants, but also in termites, birds, and other animals. If ants are just communicating one to one with no overarching supervision, then why do we see mass organization, sudden bursts of nest building or food foraging interspersed with quiet times, synchronous behavior between many individuals?

What appears to look like good evidence for social organization at a high level may yet again reduce to some plain rules. Some of the synchronicity is thought to be caused by those feedback loops we saw in operation as the ants built their nest. A trigger from a few ants ripples through the colony as they communicate with one another. Add in some programmed tendencies, like a natural period of quiescence after a sudden bout of activity, a sort of rest period not uncommon in many animals, and distinctive patterns of behavior can quickly appear to engulf the whole population. These phenomena require no superintendent to coordinate and watch over them, but rather they emerge from the self-organizing behavior of populations in communication at the individual level.

In seeing our capacity to describe ant behavior using equations, we are tempted to think that this is the whole story. Of course, ants are not mere atoms of a gas. An ant is made up of a quarter of a million neurons, the cells that transmit electrical information in our brains and in the nervous systems of other animals, including the tiniest insects. Like a miniature computer, an ant is not a mere passive observer of the world around it, like a small atom of gas bouncing and colliding with other atoms. It has oddities of behavior, perhaps molded by the ant that fell on it early in the day or the number of ants it was with earlier. And alongside that behavior, new calculations are constantly being made. The number of ants it bumps into in a given time allows it to estimate the sum of other ants nearby and so modify its conduct. Even the concentration of carbon dioxide, the gas exhaled by other ants, provides a measure of the density of ants in any part of the nest, feeding into that mini calculating machine to make it redirect its action. Proactively, ants can respond to many cues being sent their way and can initiate new behaviors that propagate through the swarm. The behaviors amplify those infinitesimal feedback loops and changes in the environment that a passive particle would ignore.

In some ways, this capacity of a living thing to respond to what is going on around it rather than merely acquiescing to perturbations in its world is a categorical difference between a living and nonliving entity. However, those reactions are still within the fold of the overarching physical principles at work. The reactions complicate the matter, but they do not put living things outside the realms of rules and principles within which they can operate—principles that we can, with enough experimental and theoretical effort, fathom.

This union between physics and biology operates beyond the imperium of insects. Far above the troglodyte lair of the ants, physicists have been attempting to unravel the mysteries of birds.

Since the ancients, humans have gazed with joy at the sight of geese gracefully winding their way across the sky in echelon or V formations, apparently coordinated and organized. Equally impressive and grander in scale are the murmurations of starlings. Sometimes thousands of birds, huddled together in a giant pulsating wave, sweep and dive in an evening sky. The self-organization of these masses attracted the attention of physicists in the 1990s; perhaps with some trepidation, the scientists launched into attempts to understand these phenomena, apparently some of the most complex in the natural world.

Hampering efforts to explain how birds organize themselves in such splendid displays was a lack of computer power to run simulations and the difficulty of getting real data. Tracking several thousand birds jostling and changing direction in three dimensions is no minor technical task. Yet advances in computer processing power, better cameras, and image-recognition software allowed people to collect some real information about bird flocking. Perhaps most surprisingly, computer gamers and filmmakers threw their efforts into the fray. Sometimes, help comes from unexpected places. Need a flock of birds in your film? You had better make them look realistic. As computer-generated sequences in blockbusters became more prevalent, so too arose the need to accurately portray birds, fish, migrating wildebeests, and a whole variety of Disney superstars. Hollywood met science.

At the core of these new attempts to simulate how birds flock are some basic assumptions about their behavior. We must establish some basal rules on how they operate. It is safe to assume that birds want to avoid collision as one condition of their behavior. Otherwise, flocking would be a bruising and messy business. They want to align their headings and stay grouped together. If they do not do that, the group will disperse, and we would quickly have individual birds heading off in random directions. We can get more complicated if we want to. We could assume that birds will try to match the speed of nearby birds, part of the strategy for staying grouped together.

Take these properties, and put them into a computer, and you can produce strikingly lifelike simulations of clusters of birds and other flying animals. So much so that the bat swarms in the film *Batman Returns* were generated using these simple algorithms.

The complexity and the subtlety of these models have been magnified in recent years with arguments and discussions over the details. Should the important rule be keeping a certain radius around each bird, or is it the number of individuals nearby that matters? How do you estimate and account for attraction and repulsion between birds, considering that they do not merely behave like particles that either collide or stay apart, but that they will avoid neighbors or try to get closer? Deciding on these sorts of intricate elements is no easy task, and the whole enterprise is made more difficult because we do not actually know what is going on in a bird's head. What calculations are really being made? A model may reproduce something realistic, but it is not based on how birds are thinking in the natural world. A scientist without a birdbrain is limited.

Like our ants, birds too are subjected to evolutionary pressures. They might want to minimize the energy they use, to conserve it for breeding. They might be in an area with a high density of predators intensifying the birds' tendency to swoop and veer to avoid being eaten. As darkness falls and their visual acuity drops, their behavior might change. And so on. Myriad environmental cues and selection pressures influence flocking. But similar to the situation with our ants, these influences seem to be just a veneer of complexity on the underlying rules that guide their patterns of behavior.

If you watch a flock of birds the way people observed ants, you can become easily convinced that one bird must be leading them. If that sort of group was a bunch of human hikers out on a ramble with no leader, chaos and misdirection would soon ensue. Just as we do for insect communities, we project the structure of our own societies onto birds and assume that the apparently organized behavior of a mass of them must require an avian superintendent to guide the flock. It feels counterintuitive to think that such organization could happen without an organizer, that disintegration of the regularity of the flock must surely occur if there is no oversight. Yet rules applied to particles in a computer show that self-organization can emerge to produce the phenomenal complexity of flocking behaviors with no head bird.

The gulf between biological behaviors and our ability to present them in physical principles, in equations, is narrowing. The infant state of our true knowledge of self-organization does not lessen the quite impressive strides made in using equations to produce realistic simulations of animal flocking, bringing us to the world of computer-generated starling murmurations. As those models are refined, no doubt the accuracy will improve and the collaboration between physics and biology will deepen as their common ground is found in one of the most ambitious programs between the two fields—to predict the behavior of populations of living things.

There is one aspect of all this that we have ignored so far. It is something physics is less able to predict, but it is singularly important in understanding why those equations work. The principles that govern flocking birds do not tell us why they do it in the first place. If you watch a vast display of starling murmurations, you are immediately enticed by the question of why. An obvious idea is that they are avoiding predators, the classic notion of safety in numbers. Faced with thousands of birds, the predator, perhaps a hungry hawk, must select one, and with such large numbers, an individual's chances of being picked off are minimal.

The problem, as keen ornithologists soon recognized, is that the birds seem to flock at the same time and place every day. Their displays often last for over thirty minutes before they settle down to roost for the night. Surely, after a few days, rather than throw off predators, this regularity would attract predators, which would quickly learn that several thousand potential meals take to the sky each evening at a particular place. Quite apart from that, the flocking behavior—every evening—seems remarkably wasteful of energy.

There is another important thing for a bird to think about other than whether it is about to be eaten. The number of birds in a flock will affect the number of available roosting sites and the available food that can go around. One advantage of taking to the air in a co-ordinated evening group has been suggested: individuals can assess the size of the flock and thereby make simple, instinctive calculations that might change their breeding behavior. How many chicks you want to have is sensibly organized around how much food and housing you have access to. By carrying out this regular census, the birds can modify their behavior to improve their individual chances of producing offspring, the ultimate arbiter of evolutionary success. This explanation of the evening flocking behavior may not require some spurious appeal to birds behaving for the benefit of the group or species, but could be driven by improvements in individual success. However, evidence for it is weak.

The real purpose of the murmurations remains something of a mystery and, like much in the natural world, may have no single answer. Instead, there may be multiple benefits to those magical displays. But our lack of insight into why they happen does not prevent us from making formidable progress in understanding how they occur and the universal rules that may shape them.

Turning our gaze from the starlings, we are struck by equally beautiful and alluring displays by larger birds, the echelon shapes of geese honking gently in their graceful journeys across the sky. Very different from the starlings' murmurations, these formations too have not been lost on physicists.

Take some hypothetical computer geese, and give them some rules, similar to those that apply to the starlings. The geese should try to stay near the closest bird and avoid collision, and they should attempt to stay where they have an unobstructed view of the way ahead. Additionally, and unlike the starlings, each individual should try to stay in the upwash of the bird ahead of it. Now this latter rule turns out to be important because one theory for why birds like geese fly in these long, strung-out formations is to save on energy. By placing yourself in the vortices of air spiraling off the wing tips of the bird in front of you, you gain the advantage of those swirling air currents curling up under your wings and giving you some lift. The notion that those fanciful formations are actually about aerodynamic efficiency in long migrations across the continents and oceans is one leading theory for why birds would do this. When you run these rules on the computer geese, they can produce extraordinarily lifelike flocks of birds that adopt all the patterns seen in the wild—echelons and V and J shapes.

But the nagging question remains: is saving energy really the reason for gander goose's geometries? In another serendipitous link between scientists and filmmakers (the acting world seems to have an attraction to flocking birds), Henri Weimerskirch, an ecologist, stumbled across a film company training a flock of great white pelicans (*Pelecanus onocrotalus*) to fly alongside microlight aircraft and powerboats so the filmmakers could get stunning footage of bird formations. Weimerskirch saw an opportunity to test the aerodynamic efficiency theory. By attaching heart-rate monitors to the birds, he showed that the birds had a heart rate 11 percent or more lower than solitary birds did, supporting the idea that the upwash in flight was enough to cut down the energy demands of flying. A small margin perhaps, but over a vast migration of hundreds or thousands of miles, such savings might make all the difference.

Because pelicans are social animals, an explanation for Weimerskirch's findings could be that the solitary birds were stressed. When the loners were deprived of their friends, perhaps their heart rates shot up, explaining why in organized groups the birds were less stressed and had lower heart rates.

However, about a decade later in Austria, a conservation group was training some other birds to follow a microlight. This time, it was not for filming, but part of an attempt to get the northern bald ibis (*Geronticus eremita*) to return to Europe by retraining them in their migration route. The conservation group hoped that once the birds had been shown the way, they would remember it and their journeys to Europe would be reestablished.

Attaching data loggers to the birds that monitored not just heart rate but also the position, speed, heading, and every wing flap, the research team from the Royal Veterinary College in the United Kingdom tested Weimerskirch's earlier findings. They too found that the birds saved energy by flying in formation, the upwash from the birds helping lift each other in their long migratory journey. More than that, it also seemed that they coordinated wing flapping to ensure that the beats were not merely a jumbled mess of flapping appendages, churning and turning the air around them, but instead synchronized to each other, apparently to minimize the possible effects of downwash, which would achieve the opposite of what the birds wanted.

In the majestic beauty of geese flying in formation in the dying light of a summer evening, we find physical principles at work: the need to stay aloft, the aerodynamic lift of a wing countering the weight of the bird, and formation flying to reduce their energy needs. Birds can follow these imperatives apparently by observing some arrestingly simple rules of engagement.

In the same way we thought about the ants, we might wonder where the individuality of birds, their personal idiosyncrasies, fit into all this. When I was eighteen, I traveled to the Canadian north to spend a month in an isolated wooden hut on the banks of the Horn River. Equipped with a marsh boat, I and three colleagues who, like me, found their way to this isolated spot through the US Fish and Wildlife Service, aimed to catch ducks and ring them as part of a survey to figure out their migration routes across North America. To the farmers living below, they were vermin, stopping off en masse to feast on corn. By establishing their routes and times of migration more clearly, we might perhaps help mitigate their damage or devise better conservation efforts. Each morning, we would buzz across the marshland with our hovercraft-like contraption to pull the ducks out of the nets we had erected across the landscape. We would then band their legs and set them free.

One thing that stuck with me, more as a point of amusement than any notable insight, was the individuality of ducks. Pull them out of the trap, and some would peck and the terrified others would freeze utterly. A few would quack uncontrollably, some would scratch, and others would sit there, humming or murmuring softly. Each animal behaved differently and had its own unique personality. I am not sure why that should surprise me. Cats and dogs have personalities, and perhaps it is rather species-ist to assume blandly that they would all look and behave the same.

I recall the Canada experience here because despite the variation in duck character, which I'm sure applies to starlings, pelicans, and northern bald ibises too, duck behavior as a collective can be predicted, and their movements and the physics behind those movements explained. The imperative a goose feels to minimize its energy expenditure while flying across continents is little affected by how it feels that morning or what its experiences have been. They are nothing more than cold calculations of energy use and aerodynamic lift. So although there may be a tendency to claim that life is highly unpredictable, that the vicissitudes of individuals make living things as far removed from atoms of a gas as they could be, these idiosyncrasies are mere tinsel on the underlying patterns that organize collections of creatures. This same idea inheres from communities of organisms right down to the biochemistry of the subatomic particles they shift around to make energy. Physics trumps individuality.

The extraordinary self-organizing patterns seen in ants and birds are to be found at lower levels of life's hierarchy as well. Although the macroscopic manifestations of these regular arrangements catch our eyes as we see them day to day, their principles unite life at all its organizational levels, an example of physical rules threaded through life.

Nothing could be further removed from the murmurations of starlings than the behavior of the slender filaments that hold together the cells that make up you, me, and the birds. Although this excursion takes me prematurely to the molecular level, a journey we will embark on as we travel down through the hierarchies of life, this brief diversion here will show the importance of self-organization as one unifying theme in the predictability of living things.

Running through and along the edges of your cells, the units from which your body is composed, are long, thin filaments. Like microscopic scaffolding, these microfilaments are made up of actin, protein tendrils stuck together in a long, winding spiral. They provide structure to the cell.

This cellular skeleton, the cytoskeleton, seems a thing of exceptional abilities. How does something made from mere spirals of a protein organize and direct so many functions of the cell? As for ants and birds, our instinctive answer is that the cytoskeleton must be controlled. To a certain degree, it is. The molecules from which the cytoskeleton is constructed are programmed by the cell's DNA. But lurking among these filaments are rules of self-organization, those same shadowy capacities that emerge in the ants and birds when we put things together at high numbers.

We can see this behavior in a laboratory. Filaments are a little easier to manipulate than flocks of birds, and in a significant paper, Volker Schaller from the Technical University in Munich and colleagues reported some simple lab experiments. They placed actin filaments on glass surfaces with some myosin, a protein that will bind to actin, and with the addition of some chemical energy in the form of ATP (adenosine triphosphate), the myosin will begin to "walk" along the filaments. In the real world, this march of myosin along actin filaments is what drives muscle contraction when you walk about or flex your arm.

Keep the actin filaments at a low concentration, and little happens. They move around haphazardly across the surface, randomly shunting this way and that. Now increase the concentration so that the actin filaments are bundled together, and an impressive transformation comes over them. They show coordinated movements.

Giant waves and vortices of filaments begin to self-organize spontaneously as the movement of every filament influences those around it. Swirling and gyrating, these structures are an elaborate mixture of both short- and long-distance interactions between the filaments. Even the computer models made by the researchers to simulate the filaments could not completely replicate the mingled mosaic of the bundles.

Other parts of the cytoskeleton show these same extraordinary powers. Tim Sanchez from the University of Brandeis, Massachusetts, and his colleagues played around with filaments of the protein tubulin. About four times the diameter of actin, or about twenty-five nanometers across, tubulin filaments also form part of the cell's skeletal structure, providing minute motorways along which molecules, essential to the cell, are transported. They organize and direct the movements of chromosomes, that is, packages of DNA, when the cell divides. When the scientists added the tubulin filaments to a dish with kinesin, a molecule that, like myosin, can crawl its way along the filaments, they too observed patterns of self-organization. Long, streaming tapestries of filaments buckled and folded, driven by the active behavior of their marching molecules.

In these astounding experiments, we see how even at the molecular level, biological entities that affect one another can spontaneously form ordered structures. Gone is the view that anything organized in life must be overseen and that this supervision, when stopped, leads to a cessation of living processes.

The rules and principles that come to the fore in a cellular filament, an ant, a bird, or even shoals of fish or migrations of wildebeests show the power of physical processes to direct and shape life at the scale of whole populations. There are common themes: the role of a critical number in transforming one type of behavior to another and the place of small, random fluctuations in dramatically driving a collection of living things into a new state. Of course, we find variety in the other rules that impose themselves on these collectives. In the air, aerodynamics, irrelevant to an ant, takes center stage in imprinting itself on a family of geese. But barring the obvious influences of other physical laws that life must also observe in different environments, is it so mad to think of the murmurations of molecules in a cell and starlings in the sky as similar?

It is easy to think that much of the complex behavior in the natural world—the murmurations of starlings, the underground dominions of ants, and the swarming of cellular filaments—is the product of something far from physics. Step out one evening, and watch the swooping and gentle swaying of thousands of starlings making an aurora-like pattern through the sky. The sight is mesmerizing, a show of such unpredictability and beauty that anyone would be forgiven for thinking this was some gift of life on Earth, something that stands above physics, something rooted in a higher order of organization. Yet within these mass organizations, there is simple order, predictability, and physical principles. Yes, there is room for chaotic behavior. The chance movement of a starling here or the collision of several there might well send the flock flowing in a new direction. It is the propagation of these small chance alterations to the whole flock that gives the whole edifice that lure of something more than physics.

It we were to travel to an alien world and observe ant-like creatures—simple things that communicate using chemicals they pass to one another—building a nest, we would see the same feedback processes giving rise to ordered communities. So too with larger creatures that may have taken to the skies of that distant planet. We might see variations between species in how they fly, for sure, but at their heart, there are physical principles that guide them in their flight, equations that order and shape societies and collections of living things. The self-organization of life shows stunning diversity entrenched in fundamental rules we might reasonably conclude are likely to be universal.

CHAPTER 3

THE PHYSICS OF THE LADYBUG

IN THE DEPARTMENT IN which I work, we offer undergraduates an option called Team Projects. In essence, you find them something interesting to work on, and they spend a semester digging around and, hopefully, in the process, learn something new.

Moving down one level of the hierarchy from populations of creatures to individuals, I thought that this general question of which laws and equations fashion life at the level of a single organism had some merit as a project. In the winter of 2016, I set my group a simple objective: spend a few months investigating the physics of a ladybug. Metaphorically take the ladybug apart bit by bit, consider every facet of its life: its airborne life in the Meadows, its perambulations on leaves, its breathing, the strength of its protective wing cases. Write down the relevant laws and equations with which that little insect is constructed and that play a dominant role in defining its day-to-day life. Tell me everything I need to know about physics to apprehend how a ladybug works. This was no small undertaking, and I was aware before they arrived in my office for the first meeting that the task I had set them probably covered an enormity of knowledge. Aerodynamics, diffusion, locomotion, thermal inertia—it was easy to casually list a mélange of physics that had at least some part to play. As I expected, the project took them on a fascinating journey into the many ways in which physical principles shape life.

A good place to start is its legs. Simple though they may look, packed into those appendages is a fascinating array of physics. Because each little leg of the insect has three joints, it can move them around in all sorts of interesting contortions. With so many degrees of freedom, the ladybug has multiple options for how it might just put its foot down. In its head, a computer is at work deciding among those possibilities. Wind speed, surface irregularities, bits of leaves, and no end of other fine details probably feed into the decision on how to move each of those six very adaptable legs.

At each step, the ladybug must ensure that it can hold on, such as when climbing vertically up the hand of a tourist, because otherwise, the insect would fall off. The ladybug, like spiders, lizards, and other beetles, has on each foot a pad covered in tiny hairs, or setae. These hairs play a role in attaching the creature to a surface. The problem the ladybug must solve is to ensure a good connection between the pads on its feet and the surface. To make the connection hold fast, it uses a very thin layer of liquid. Secrete a film of fluid under your foot, and at the small scale of insects, the fluid produces a huge adhesion force by capillary action and the viscosity of the fluid. By producing a liquid in this way, the ladybug fills in irregularities on the surface, essentially making the surface behave as if it were flat. The layer of fluid is thin and keeps the friction high enough to prevent the insect from slipping down the vertical face.

By combining all this newfound knowledge about the ladybug's foot and the behavior of fluids, we can even write an equation for the total adhesive force that the ladybug leg can produce. With it, we can predict its ability to master and own the terrains that constitute its little world. That memorable equation is:

$$F_{(adhesion)} = 2\pi\gamma R + \pi\gamma(2\cos\theta / h - 1 / R)R^2 + dh / dt \, 3\pi\eta \, R^4 / 2h^3$$

where $F_{(adhesion)}$ is the adhesive force. γ is the surface tension of the fluid under the foot. R is the radius of the foot, considered a simple disk, although this can be made more complicated and realistic in shape. The value η is the viscosity of the fluid under the foot, h is the distance between the foot and the surface, that is, the thickness of the liquid layer, and t is time. The first term is the surface tension, the second one the Laplace pressure, and the last term the viscous forces.

But in walking across its varied and unpredictable terrain, the ladybug has a little problem. The legs must be rigid or the insect will collapse, but we want malleable feet that can move flexibly across those surface irregularities. To achieve this, the leg contains a protein called *resilin*, an elastic biological polymer that enables fleas to jump and other insects to do contortions when they need to. From the top of the ladybug leg to its feet, a gradient of resilin has been discovered, with more of the substance nearer the feet, where elasticity is needed, and less toward the top of the leg, where rigidity is preferred. The top of the leg contains more of the tough chitin that makes up the rest of the insect's exoskeleton, increasing the Young's modulus, a measure of the stiffness of a material. Here we find that the physics of material properties have evolved to suit the needs of the perambulatory part of the insect's lifestyle.

The forces that attach the foot to the ground are impressive, but we also must be able to pull the legs back off the surface again. Otherwise, our insect is rigidly transfixed to one location, pulling and tugging in vain against the laws of physics. Simple equations can again define the energy needed (W) to pull the feet away from these adhesive forces:

$$W = F^2 N_A l g(\theta) \,/\, 2\pi r^2 E$$

where N_A is the density of setae on the feet, l is the length of the setae, E is the Young's modulus of the setae, a measure of their tendency to deform, and r is the radius of the seta. The term $g(\theta)$ describes the angle of the setae to the surface and is given by $g(\theta) = \sin\theta[4/3(l/r)^2\cos^2\theta + \sin^2\theta]$.

Now the story is not over, because the insect, to really hold on, wants to have lots of those hairs, but not too many. If it has too many, they will all get stuck together. They must be far enough away from one another that the forces of attraction between them are not too great, but close enough to maximize the number on the feet. That theoretical maximum density of the setae is given by:

$$maximum \ N_A \leq 9\pi^2 r^8 E^2 / 64 F^2 l^6$$

where F is the adhesive force of one seta.

To really pack these setae in, the insect can evolve some modifications. If the setae are sticky on only one side, the chances that they will get stuck together can be minimized. Protrusions and nodules on the foot further keep them separated.

The insect must evolve to ensure that all the solutions to these equations are optimized in a way so that the feet can hold on tight but be pulled away as the ladybug moves across the ground. The setae, those hairs on its feet, do exactly this, a superb and exquisite example of the honing of biological form by simple physical principles that can be expressed in equations. It is no surprise, then, that hairy pads on insect feet have evolved time and time again completely independently. Those physical limitations are unyielding in their grip, and yet the solutions to them in the living form are limited, bounded by some straightforward outcomes in the insect. Convergent evolution is here exposed in all its glory as the mere channeling of biological form into similar outcomes determined by physical principles. All insects converge on some relationships, complex and fascinating in their totality and yet reducible to understandable simplicity.

It is not immediately obvious to you and me that a small insect would use these cunning means to walk across surfaces, up walls and even upside down on leaves and ceilings. Try covering your feet in a thin layer of water and then walking vertically up the outside wall of your house, and you will soon be sorely disappointed. At human scale, the forces of gravity dominate, yanking you remorselessly off the side of the building before you have had time to take a single step forward. For the ladybug, a thing about seventy-five thousand times less massive than you and me, molecular forces dominate. Surface tension, capillary action, van der Waals forces, and others. All these forces that yank and pull molecules together shape a ladybug's universe and lead inevitably to the little tricks of using the strength of these molecular forces to stick to walls. Not that gravity is an irrelevance to ladybugs. Fall off a leaf or a wall, and they will fall as surely as you and I, although slower and with less damage. Gravity is inescapable. Yet at the scale of most insects, it takes back stage as molecular forces come to the fore.

As some doors open, others close. The ladybug may well use thin layers of a liquid to climb a wall, but those same molecular forces, those identical laws, bring disadvantages too. When you and I and other large animals need to wash, we enjoy a shower, a bath, or the nearest watering hole or pond. Climb out, and most of that water drains away under gravity, with just a thin layer and a few droplets of water remaining. A dog will shake them off and you will use a towel, and if neither of these possibilities are available to you, well, you can just wait for the water to evaporate.

Ladybugs need be more careful, though. Even a single drop of water will relentlessly cling on, those surface-tension forces too much of a match even for its strong little legs as they attempt to push it away. For something smaller, like an ant, the insect may be completely consumed by such a droplet, trapped inside, the molecular attraction between the water molecules on the surface of the droplet imprisoning it in a water cage, a surface-tension prison. For these reasons, many insects, particularly small ones, dry-clean themselves, scraping dust and dirt from their bodies with their tough legs while avoiding the ensnaring promise of water.

Some of these observations—that humans cannot walk vertically up walls but flies and ladybugs can—may seem so obvious and taken for granted in our everyday experience that they are hardly worth remarking on. But they define two worlds: the physics operating in the ladybug's domain and the physics you and I must observe. The same physics, for sure, but different forces dominating in each realm. Yet this physics explains so much about the form and shape of creatures that inhabit these different scales. Nothing is trivial about the design of a ladybug's legs and the way it must wash itself or the limits of locomotion in a human or a gazelle. However, we can unravel these limits and possibilities not by exploring contingency, in other words, the historical quirks of life's evolution that could be very different if evolution were rerun, but only by investigating the fundamental physics that rules how these organisms operate.

Ladybugs can do much more than walk across obstacles.

Conspicuous among the ladybug's collage of skills are minuscule wings, which are packed under their wing cases, the elytra. Gossamer thin, a mere half micron thick, they are cleverly folded in thirds, protected by the shiny, hard wing cases. If the insect is distracted by a predator or an Edinburgh tourist, in under half a second, those wings unfold into their articulated state and can carry our little insect high into the sky, if it so wishes, over a kilometer up, traveling at speeds of up to sixty kilometers an hour.

The ladybug wing is no mere simple fixed structure, like an airplane wing. This is a flapping contraption that, fully expanded, is four times the length of its body. Attached to the body by hinges, the wing swings backward and forward in the horizontal plane, the muscles in the body and the veins in the wing acting as levers to push the wing through the air, generating lift from this airfoil as it launches itself skyward. Running along the front of the wing is a reinforced vein that provides strength and stability during collisions with raindrops and other unexpected objects.

Although the ladybug wing has great flexibility, the muscles at its hinges and the veins within the wing allow the wings to change shape and twist in response to changing air patterns and wind. However, on the face of it, even these impressive structures seem unlikely to hold such a rotund little ball aloft.

Trying to work out how insects fly has consumed the efforts of many entomologists. Rapid advances in computer modeling, helped along by high-speed photography of insects in motion, have allowed researchers to explore the exquisite refinements that allow insects to exploit every subtlety and possibility that emerges from the physics of aerodynamics.

Those wings can do all sorts of clever maneuvers to gain lift. Within their apparently chaotic and mad swishing are intricate coordinated changes, one with the mysterious name of *clap and fling*. As the wings are pushed into their backstroke, they are clapped together, forcing air out from between them, thrusting the ladybug forward. The wings are flung apart as they begin their front stroke, and the air that rushes in to fill the gap enhances circulation over the wing surface, improving lift. Clap and fling has its problems, not the least the tendency to damage the wing with such violent and abrupt adjustments. Instead, increasing the wingspan or the frequency of wing beats can augment the lift the insect can achieve.

Not surprisingly, with this knowledge in hand, we can reduce the wings of the ladybugs to equations, and once we do that, we can calculate the lift and power generated in these appendages. By considering the forces around the wing, the angular velocity, and the moments of inertia, we can reduce the flight of the ladybug to a number as simple as about thirty watts per kilogram of power produced by those appendages.

As the creature touches down, it must now cover up and protect those wings lest they are damaged. Inward they flip, beneath the tough wing cases, safe from harm. The delicate wings are now hidden beneath the two shells, which are cleverly evolved to slot together along the middle in a tongue-and-groove system, similar to the slating in floorboards.

Nature needs a material to build the insect from, including the wing cases. Its solution, chitin, is a robust sugar material about ten times less strong than steel, but about ten times stronger than the material from which hair is made: keratin (a material made of protein). Unlike the legendary strength of spider's silk, we do not need super durability here; we just need something that will do the job of protecting the wings and the rest of the ladybug's vulnerable parts, including its head.

Throughout the ladybug's armor, chitin is woven. And sometimes, such as in the antennae and legs, we find resilin mixed in for some flexibility.

As our ladybug bashes and crashes its way through life, the successful ones that get through to reproductive age without having their wings written off will be those that can withstand all those collisions. The severity of collisions can be calculated with equations such as the Head Injury Criterion (HIC), a practical equation used to work out how effective your bicycle helmet is at protecting your own head:

$$\text{HIC} = (t_2 - t_1)\left[1/(t_2 - t_1)\int_{t_1}^{t_2} a(t)dt\right]^{2.5}$$

where t_2 and t_1 are times, and $a(t)$ is the acceleration in the collision.

Albeit more complex in a shifting, dodging ladybug, the strength of the chitin and the thickness of the wing cases must withstand the accelerations of the collision to keep the wings intact. Ultimately, those forces are fashioned by physical properties.

Now, chitin is a translucent material, and the wing cases, as well as covering the wings, must perform some other functions. Well, yes, attracting a mate is important, as is warding off predators. Impregnated onto the wing cases' otherwise lackluster colors are reds, blacks, and yellows, themselves a mere part of the palette of colors from iridescent reds, greens, and golds found across the insect world.

Somehow, we must color those cases in a way that makes sense. Random sheens and splashes might make a nice scene for the artistic observer, but on our ladybug, as with other insects, we want patterns, dots, for example, in particular places. Patterns in animals, including dots on insects, perform numerous roles in camouflage, discouraging predators and attracting mates.

It was Alan Turing, a physicist and a computer genius, who first proposed how this coloring might be done. Imagine a cell making a colorful pigment in our newly begat ladybug. Slightly further away, another cell might be producing an inhibitor that prevents that pigment from being made. As these pigments and their inhibitors diffuse through the cells of the developing insect, we end up with gradients. At just the right places, the pigments can be produced and maybe a little black spot will appear. In other places, the pigments are suppressed and maybe the insect goes red. By varying the range over which these different activators and inhibitors can work, all manner of patterns can be produced in the final product. Through these simple rules, we can make ladybug dots, leopard spots, fish stripes, and Dalmatian dogs. And these gradients can be written in equations, for example:

$$\delta a/\delta t = F(a,h) - d_a a + D_a \Delta a$$

$$\delta h/\delta t = G(a,h) - d_h h + D_h \Delta h$$

where t is time and a and h are the concentration of activator and inhibitor, respectively. The first term describes the production of the chemical, the second term the loss due to degradation, and the last term the diffusion of the substance. There are many variations on these equations.

Turing patterns are common in nature, and of course, in reality they are much more complex than mere activators and inhibitors spreading out from some cells. Multiple chemicals play roles, and other metabolic interactions add complexity to the final design. In the developing embryo of many animals, genetic control and regulation—rather than simple gradients of chemicals—take center stage. But the essential idea that Turing put forth about how the interacting gradients of chemicals produce patterns provides a clever basis for understanding how the ladybug got its spots. Although Turing's simple model has frequently been superseded by more-complex knowledge of genetics, his work was a gallant effort to explain complex biological phenomena with simple physical principles and was one of the early attempts to unite physicists and biologists.

Having warded off a predator or maybe spotted a mate with its dotty body, the ladybug must take to the air. To launch itself from the awakening tourist in the Meadows, the insect must be warm enough to flap its wings. The insect is an ectotherm, gaining most of its warmth from the environment and not from its own metabolism, unlike you and me—we are members of the endotherm club, warm-blooded animals that regulate their own temperature. The ladybug must rely on its surroundings to keep warm. That is a problem for our friend because below a temperature of 13°C, it will seize up from the cold. Thus, an essential part of a ladybug's life is to bask in the sun, an activity I'm sure many of us would like to justify during our working hours as a physiological requirement.

By sitting with its wing cases facing the sun, the ladybug can absorb solar radiation through its chitinous covering and therefore raise its temperature. Some of that radiation will be reflected from the shell's surface. Those bright, attractive colors come with a cost— they are much more likely to reflect the very radiation needed to keep warm. Indeed, dark ladybugs are known to be more efficient at collecting valuable solar energy than are light-colored ones.

In you and me, some of our heat is carried away in evaporation. The water in our sweat carries away the energy as it is vaporized from the surface of our skin. The very purpose of sweating is to prevent overheating. Luckily, for the ladybug, little evaporation can occur under its thick wing cases, so it is spared an unnecessary loss of heat. However, some heat will be conducted away from its body through the air.

Carefully and systematically, we can add, as in a ledger, all the sources and losses of heat in a ladybug. We can think about the sun's energy coming in and how much is reflected away. We can consider the small amount of warmth coming from the insect's metabolism and making its way from the inside out. We might factor in that thin layer of air between the body and the wing case—a layer that acts as an insulator, slowing heat loss. We could think about the air blowing over the creature, which will carry off some of the heat.

When we add, subtract, divide, and multiply the various parts of this tapestry of heat fluxing back and forth, we can arrive at an equation for the temperature of a ladybug (T_b). The little equation captures the difference between life and death for the tiny beast:

$$T_b = T_r + tQ_sR_b(R_r - R_b)/kR_r$$

where T_r is the temperature of the wing cases, t is the transmittance of the wing case, Q_s is the incident energy on the ladybug, R_b is the radius of the ladybug body, R_r is the radius of its shell, and k is the thermal conductivity of the air between the body and the wing case.

As the sun on the Meadows sets in the evening twilight or the cold chill of winter descends on this beautiful part of the city, sending townsfolk and tourists into homes and cafés, the little ladybug has no such respite. T_b drops into the red. But it has other tricks. Almost imperceptible to you and me, ladybugs can shiver. In so doing, they burn up more energy and produce heat, offsetting the loss of heat through their shells and keeping their temperature high enough to move around.

However, soon, the shivers of a ladybug are no match for the winter and it must go into hibernation. For up to nine months, the temperatures of our Scottish city threaten to force this minuscule critter into torpor. Off it heads, joining forces with friends across the city to find some old leaves, soil, moss, or other insulating material, and together they will huddle, surrendering to the cold, allowing their bodies to cease activity. Walking and flying are now difficult, and within their snuggled-up family of friends, they are protected from predators that would otherwise take advantage of their sluggish form.

In the dead of winter, at the apex of this long sleep, temperatures may drop below freezing and now the ladybugs have another problem. The cold temperatures that made them listless are a mere picnic against the possibility of temperatures that threaten to freeze them solid, raising the specter of ice crystals, sharp and nasty, crashing through cell membranes and causing irreparable damage.

The beetles must now call into action another equation:

$$\Delta T_f = K_f m$$

Add salts or other antifreeze substances to the body, and the freezing point of water can be lowered. The drop in freezing point (ΔT_f) is established simply by the product of the *freezing-point-depression constant* of a chemical (K_f) and m, the molality of a solution.

Add a gram of the chemical glycerol to two grams of water, and the solution now freezes at about −10°C. By synthesizing these sorts of compounds in their blood, our ladybugs can withstand the precipitous drop in temperatures, preventing ice crystals from forming, and come out on the other side of winter little worse for wear.

In these equations, we have a marvelous illustration of how life must deal with the unrelenting and ineluctable principles of physics, but also how, through evolutionary inventions, natural phenomena described by different equations are incorporated into living things to their advantage. The temperature of the ladybug is a concoction of interacting terms settled in final mathematical form by the heat terms that come and go from its body. In the darkening sky, the outcome for the ladybug is a foregone conclusion. The creature will begin to chill. In some part, it can take action against this heat loss by shivering. The simple process of burning more food to make heat will offset some of that cooling. The innovative bug can do more. Legs and wings, the equations of locomotion and aerodynamics, allow it to take action that is more drastic. It may crawl into leaves or take to the air in search of a place to hide. Buried in leaf litter, it has succumbed and accepted the laws of physics, capitulated to the inevitable lethargy that must overcome it. But evolution is no passive onlooker.

Experiments, driven by mutation, explore available physical principles for possible solutions that will produce organisms capable of more successfully reproducing. They define and fashion future populations that are more robust and successful. In the history of insect biology, the production of compounds such as glycerol to push the freezing point down was one such innovation that employed the freezing-point-depression equation to expand the temperature range of ladybugs, to push their lower limits downward and into the realms of the otherwise-frozen world.

In this equation for freezing-point depression, we have an exquisite example of how variant offspring in a population will explore new physical relationships hitherto unrepresented in the living forms of their ancestors. Physical principles stumbled on by new variants that enhance their chances of reproducing will be selected for and passed to new generations. In this way, life explores the universe of these principles, which are given expression in equations, and incorporates them within the living form.

Simple physical relationships have a role to play not merely in movement and temperature regulation, but also in the gases that animals need to survive. The ladybug's ability to move, heat itself, and reproduce depends crucially on its capacity to get that most vital of gases needed by most animals: oxygen. In you and me, the lungs pumping air in and out provide us with a continuous source of the gas. Even the fish, in their watery world, flow water through their gills, extracting the oxygen essential for gathering energy; like us, they use this gas to burn organic food.

Like all insects, the ladybug does not have lungs. Instead, cylinders run through its body like a network of life-giving tubes. These trachea, which themselves connect to tinier tracheoles, traverse its insides. Air travels through these hollow lengths, delivering oxygen to the insect's innards. The network is so comprehensive that the oxygen can be delivered to within micrometers of where it is needed. However, this transport depends on the passive process of diffusion, the movement of atoms or molecules from a place of high concentration to one of lower concentration.

The relative simplicity of insect breathing lends itself to uncomplicated equations that deal with this process of diffusion. The time it takes (t) for a molecule to travel a given distance (x) can be worked out using the equation:

$$t = x^2 \, / \, 2D$$

where D is the diffusion coefficient, a measure of how quickly the molecule can move in a given medium.

This simple equation tells us that the time it takes for a gas to travel a given distance is related to the distance it travels squared. Double the distance you want your oxygen to go, and you must wait not just twice as long, but four times the time. And it just gets worse as the numbers are increased. That is why insects are ultimately limited by diffusion.

The gases going into the ladybug diffuse at a rate that depends on the concentrations inside the body. Adolf Fick, a physiologist, was a pioneer in getting to grips with diffusion. In the nineteenth century, he worked out a great deal of what we understand, including his well-known Fick's first law, which tells us the amount of a gas, like oxygen, that will flow at any given time. His equation is as follows:

$$J = -D \; dC/dx$$

where J is the flux, D is the diffusion coefficient (again), and dC/dx is the rate of change of concentration of the gas over a given distance. This equation allows us to work out how much oxygen will pass into the insect's body.

These equations, applied to insects, tell us a very simple thing: insects are limited in size. Get too big, and you cannot get enough oxygen into your middle in a sensible time without a very cumbersome network of trachea. The distance of diffusion is not the only problem. As oxygen diffuses in, it gets used up before it can reach the deep interior of the animal. This is one explanation for why there are no ants or beetles the size of elephants.

Working out the largest possible size of insects is difficult since complicating our so-far elementary analysis is a snag: some insects can actively pump air in and out of their bodies by moving their abdomens. In this way, convection, which is the movement of air driven by pressure gradients, can more forcefully be brought into play, allowing many insects, particularly largish ones like cockroaches, to actively pump in oxygen. But even with this aid, insects still have a limited mass they can achieve through the simple networks of trachea.

Insects can get large, but not nearly as impressive as the most sizable mammals or extinct reptiles, the dinosaurs. The largest recorded insect is a specimen of weta, *Deinacrida heteracantha*, a cricket-like insect that lives in New Zealand. The 71-gram giant comes in alongside the goliath beetles that routinely weigh in at over 50 grams. But compare these Godzillas of the insect world to the 140,000-kilogram mass of a blue whale.

Peer back through time, and you see something very odd. About three hundred million years ago, bigger insects, much bigger, inhabited the world. Buzzing through the skies of the rich Carboniferous forests—enormous swampy expanses of trees that would ultimately provide us with coal—were immense dragonflies. The now extinct *Meganeura* had a wingspan of well over half a meter, and creeping its way through the undergrowth was the terrifying *Arthropleura*, a millipede that grew to over two and a half meters. What happened at this time? Was evolution on a meander through a random experiment in giant insects? Coincidently, at the same time, the oxygen levels in the atmosphere rose to about 35 percent, compared with today's 21 percent. That elevated oxygen may well have played a role in allowing for giant insects. As the oxygen levels rose, more oxygen could effectively diffuse into larger insects, allowing the body size to increase.

Like all ideas, sometimes a beautiful story can be ruined by hard facts. Oxygen has a more complicated role in life than merely influencing how big something gets because of diffusion. Higher oxygen levels can be toxic, producing free radicals that must be quenched lest they attack key biological molecules. Insects might have evolved to grow bigger to minimize these effects of oxygen in the insect body by reducing the amount of oxygen that diffuses into the interior of the animal. Larger insects have other problems, like their need for more food and the possibility of breakage in their exoskeleton. These factors also play into the overall constraints on insect size.

Nevertheless, however incomplete our knowledge of the ancient landscape for insects and their atmospheric environment, we must conclude that physical principles strongly shape the forms of insects and their maximum size. Therefore, ultimately, in comparing them with reptiles and mammals, we can see how the architecture of an animal is severely limited. We can argue for contingency and chance in molding the fine details of insects, but when confronted with the challenge of their ultimate limits, we must return to basic physics.

To find its way to its protected group of companions under the moss pile or the autumn leaves blown up against rows of the

Meadows' trees, the ladybug must be able to sense its world, and on its head are a veritable complexity of sensors to do just this. The ladybug has two eyes. Unlike in you and me, these eyes are not one large lens capturing light and transferring it to many receptors beneath, but, like all insects, the ladybug has a compound eye. In a giant cluster of minuscule individual lenses, called *ommatidia*, light is captured by each one from a different part of the sky. The size of these little lenses is limited. Naturally, the ladybug would like to have as many as possible. The more there are, the higher the resolution that can be achieved. In the other words, the more detail can be collected from the world around it. The angular size of each lens (θ) is given by the simple equation:

$$\theta = ad/r$$

where a is the angular field of view of a row of lenses, d is the diameter of the individual lens, and r is the length of a row of lenses.

If we pack in more lenses, we can certainly collect more information about the world. But a new problem emerges, since small lenses become subject to diffraction, the process that causes light to be slightly bent and distorted, causing interference. The eye becomes useless. The angle (θ_d) below which these effects begin to disrupt the ladybug's vision can be calculated using the wavelength of light (λ):

$$\theta_d = 1.2\lambda/d$$

Here again we find a trade-off in physical principles up against one another, doing battle in the evolutionary struggle. Make the individual lenses smaller, and you can fit more of them into the eye and you can see more of the world. But make them too small, and the physics of light behavior renders them ineffective. The evolutionary process is restricted by intersecting principles that forge and hammer its products into predictable and narrow forms.

The reception of light and colors by insect eyes is a field unto itself. Different ommatidia have receptors for blue, green, and, in many insects, UV light, giving them access to a region of the electromagnetic spectrum that you and I cannot see. In some flying insects, receptors adapted for UV and blue preferentially face toward the sky, perhaps to aid navigation. In the visual capacities of insects, and indeed all animals, the requirements for the physical detection of different regions of the electromagnetic spectrum meet biology.

In this diminutive ladybug, we have toured and explored just one example of the irrefragable and deep links between evolution and physical processes. After several months of study, the group of students whom I had tasked with exploring the physics of a ladybug had probed just a few principles in a report more than forty pages long. We have not even scrutinized the antennae, packed full of sensilla that can detect chemicals, physically feel their surroundings, sense air speed while they are flying, and, in some insects, pick up sound. Each capability has a set of equations we could list and explore in depth. We have not talked about the mandibles and the mechanics of the mouth used for snipping leaves and crushing food, the processes themselves a concatenation of equations and forces all of which must converge to provide the ladybug with sustenance. The digestion and absorption of food carries us down another path, where diffusion, osmosis, and friction all interplay with other forces to define how well our insect can gain the energy and nutrients it must acquire to grow and reproduce. And what about the insect blood, the hemolymph, that circulates through its teeny vessels to bring vital nutrients to the cells and remove waste? What of the physics of muscle function, the storage of energy, or the details of the tegument, the outer layer of the insect? And the physics of ladybug reproduction? The eggs, their development, and the larval stages of insects? I suspect three or more years of research would be needed to do this task true justice. All these questions lie beyond the scope of this book, but the few efforts we have embarked on in this short exploration illustrate the conclusions we must draw.

The ladybug is a remarkably complex thing, and packed into this machine, whose mass is a million billion billion billionth of the Sun's, are many more physical principles than those that define the structure and evolution of a star.

Those physical principles are not discrete, doing their own thing, but are all interwoven. In the evolutionary process, natural selection operates on each living thing to remove the variants in which the mosaic of principles given expression in them are not sufficiently optimized to allow them to achieve reproduction.

In surviving collisions that would tear and otherwise damage its micron-thin wings, the ladybug must grow thick wing cases that can withstand the unpredictability and knocks of the outdoor life. Yet if they are too thick, the greater weight of the insect diminishes its ability to fly and to flick its wings into action quickly to escape a predator. In this conundrum, the Young's modulus of its material of choice, chitin, must come face-to-face with the equations that define its aerodynamic life. The relationships that describe the strength of chitin will themselves change how that material absorbs heat, linking the equation for the temperature of the ladybug to the effectiveness in surviving collisions.

We can imagine a giant sheet of paper with many hundreds of equations written down, curved arrows running hither and thither showing how the terms or solutions of one equation influence another. Feedback processes abound as this enormous network of equations shifts and changes, each minuscule alteration in one equation, like "the wave" in a stadium, rippling through the rest. This is life. Mutations alter the solutions to some equations, add new ones, remove others. Natural selection takes the interwoven whole of this mathematical olio and subjects it to the environment. Those tapestries of physics that are manifested in ladybugs that successfully reproduce move on to new experiments. Those that do not are removed.

A fascinating challenge would be to create a ladybug in a computer with as many physical principles that can be described in equations as possible. Beyond the cursory effort here, we might delve deep down into the genetic code to add mutations and errors in the code. At the higher levels, we might produce populations of ladybugs, simulating their gathering multitudes in the cold. Aside from its scientific use in investigating in more detail the possibility of reducing an entire multicellular animal to physical principles manifested in equations, it would be a profoundly good way to deepen our efforts to understand the various forces and possibilities that shape the living form. Such efforts would take us further into the realms of predictive capability, a fundamental characteristic of science.

Reducing life to a set of equations may provide an effective means to link genetics with physics. Consider the temperature of our ladybug. Each term in the thermal equation that gives us its temperature can be considered to be controlled by a gene, a set of genes, or the emergent properties that result from the products of many genes. The solar radiation lost on the surface of the insect depends on how much is reflected away, and the amount of reflection depends on the surface's shininess, itself a product of genes and developmental pathways that decide the surface characteristics of the wing cases. Some radiation may be scattered away if the surface is rough, again controlled by genes that influence how the wing cases are fabricated. The quantity of radiation lost from the body of the ladybug will depend on the thickness of the wing cases, which is determined by the genes that control their development. And so on.

We must be careful not to reify equations. They do not exist as physical entities; they merely express relationships between different variables. However, an equation that defines a characteristic that we know helps an organism reach reproductive age, such as its thermal balance, can be thought of as a way of coalescing various physical features of an organism. And each of these features makes up the term of an equation and might be assigned to a specific gene or set of genes or the interactions of their products.

By linking the change in an equation's term to genes and their ultimate pathways (simplistically, say, a gene or genes determining ladybug wing case thickness that could replace the thickness term in the thermal balance equation), we may even express variables in an equation in terms of the activity of genes. In this way, we truly integrate physical relationships and properties in the macroscopic world with the genome and the pathways that result from it.

The influence of different environments on these genetic pathways in the whole organism might be a further variable we could add to specify how the environment influences the solution to given equations. Thus we would be incorporating processes that operate from the top down as well as the role of the genes in determining the structure of an organism from the bottom up. In essence, an equation provides a useful means to establish which characteristics of an organism should be considered as a whole system, the terms of which come together to influence an important property that bears on its ability to reach reproductive age. Many genes are involved in more than one process, and the complexity of developmental processes assures us that linking genetics with physics in this way would be a heady ambition, made difficult by the fact that, for many processes, there is no simple link between a single gene and a phenotypic characteristic. Nevertheless, this evolutionary physics or physical genetics approach, however one likes to think of it, may provide just one useful way to encapsulate adaptations and evolutionary changes in quantitative, physically circumscribed terms.

Throughout this tantalizing foray, we have also seen glimpses into common reasons for evolutionary convergence, the reasons why organisms have analogous structures. *Evolutionary convergence* is often just another way of saying, albeit more efficiently, similarity caused by the laws of physics. In their sticky, hairy feet, the shape of their wings, and the thickness and the color of their wing cases, the ladybugs show us that the simple equations and mathematical relationships that sculpt their curves are imposed on all insects. The network of equations will always lead to modifications. A larger wing here has a ripple effect on the wing case or the size of the legs there. The color of a beetle here affects thermoregulation or hibernation habits there. Through these small alterations forced on insects by their predators, food, or homes, the vast medley of insect life on Earth is produced. Yet through all this detail, the enduring equations of life channel evolution in narrow ways, bountiful, beautiful, and dominant in the phenomenon of life.

CHAPTER 4

ALL CREATURES GREAT AND SMALL

OUR EXCURSION INTO THE physics of ladybugs has revealed much about why creatures look like they do. But ladybugs are only one type of insect, and we might wonder about the rest of life on Earth. Since Darwin's landmark insights, evolutionary biology has taken a particular interest in whole creatures, from finches to fish. Like ladybugs, do all these creatures great and small show inklings of physics, and can our ladybug provide us with a foundation to learn anything more general about the link between evolutionary biology and physics? Continuing to look at this level of life—how whole organisms are shaped—we might explore how physics informs our understanding of the rest of the planetary zoo.

The fields of evolutionary biology and physics do not look, at first glance, like natural twins. However, there seems to be no contradiction between the idea that physical rules drive life into narrowly circumscribed forms and the modern view of the role of evolution and biological development in shaping form, just as ladybugs illustrate. Physics explains much about why living things look like they do; evolutionary biology provides much of the explanation about how they become like they are. Together they constitute a complete picture. A beautiful way to reveal the general harmony between physics and evolution at the scale of whole creatures is to continue to explore evolutionary convergence.

Evolutionary convergence is rampant in the biosphere, far beyond the construction of ladybugs and other insects. Let's visit one of
my favorite animals (I know not why), the mole.

Wherever they live, moles have a fairly simple objective in life—
to burrow underground, build a nest, and have offspring. Their subterranean lifestyle demands some basic biological features, many of
which are underpinned by a straightforward equation in physics—
pressure is equal to force divided by area:

$$P = F/A$$

The mole must burrow with enough pressure to push the soil out
of the way and make progress in its desire to build a tunnel or a nest.
The equation is self-explanatory—the more force you exert per given
area of surface, the more pressure you apply. For the mole, the outcome is very simple: either the pressure is greater than the cohesion
of the dirt it seeks to burrow through, in which case the material in
front of it can be displaced, or it is not. If, day after day, the pressure
persistently is not high enough, then the mole cannot dig or it will
get accidentally buried. Either way, those moley genes will not be
passed on to offspring. $P = F/A$ matters to a mole.

There is therefore a powerful and uncompromising selection
pressure to push that muck apart to make an underground mole
house, and that selection pressure results in certain predictable features. Moles have short but wide stubby front feet that minimize their
cross-sectional area, meaning that ultimately the pressure that can be
applied in front of the animal is maximized. Those paddle-like paws
simultaneously enable the mole to displace large amounts of soil. It
is a compromise between large appendages that get in the way and
increase your cross-sectional area and ones that are sizable enough to
scrape lots of dirt out of the way.

At the end of the short and powerful appendages are tough nails
that enhance the mole's ability to dislodge and shift the world around
it. It even has an extra thumb, an example of a polydactyl paw, which
helps it burrow. To improve the effectiveness of its forward move-

ments, the mole is a slender shape—a large, fat, cube-shaped animal will, all things being equal, exert less pressure on the soil, because the force driving it forward is spread over a larger area.

I have simplified somewhat. The mole's underground existence begets other adaptations besides. It has a high tolerance to the carbon dioxide that builds up in the subterranean environment as it breathes. In its blood is a type of hemoglobin, the protein that binds oxygen and transports it around the body, with a very high affinity for oxygen, allowing it to subsist at concentrations of carbon dioxide that would become asphyxiating to us. So $P = F/A$ is not everything, but even these other adaptations turn out to be driven by other laws that for the time being I will bypass.

The mole is an engineering solution to the compromises needed to effectively shift soil by maximizing the force applied over a small area. It is an organic manifestation of $P = F/A$.

As an evolutionary consequence of this law, moles look the same regardless of their provenance. The *Talpidae*, the true moles, which live in Europe, North America, and Asia, appear similar to Australian moles, or the marsupial moles, which are more closely related to kangaroos and koalas. Moleness is a product of physics, and no matter whether you are a small furry mammal burrowing in the dirt of Edinburgh or you are burrowing in the dusty plains of the Australian outback, $P = F/A$ mandates that you end up looking the same.

I chose the moles because their very specific lifestyle, which is dominated by the need to burrow, brings one law of physics to the forefront of their lives. Their lifestyle demonstrates that evolutionary possibilities, in this case the options for the form of a small burrowing animal, greatly constrain the potential solutions available in this process of convergent evolution. Many other animals that burrow display the same general design.

Even where two animals seem to have converged on the same shape not because of some physical feature of their environment but because of the influence of other life forms (maybe they are sleek and slender to escape predators), physical principles ultimately lie at the heart of those similarities. Perhaps the animals have bigger muscles to produce greater acceleration or better eyes that are more effective at collecting light with which to see predators.

Nothing is magical or strange about convergent evolution any more than there is something uncanny about the fact that when both liquid lithium and liquid water are heated, both substances turn into a gas. This latter observation is not bizarre cosmic coincidence; rather, it is just physical processes, the consequence of adding energy to the molecules or atoms of a liquid to overcome their forces of attraction and cause them to dissipate into gas. This is the same with convergent evolution; very often it results when similar physical principles operating on different organic forms drive those forms toward resemblance.

Often in convergence, more than one law may be at work, complicating our ability to identify a straightforward equation. Explore enough of the workings of any creature, such as ladybugs or moles, and you will find a multiplicity of laws relevant to their survival. If we do not comprehensively grasp the biology or ecology of an organism, we may well not even be able to identify the relevant laws. There may be more than one overall solution to the laws that operate or are exhibited in the organism. However, the ubiquity of convergent evolution across the biosphere and the vast number of examples show that solutions are not limitless, but usually rather few.

To some people, there may be something strange about the apparent ability of life to navigate the vast landscape of possibilities to arrive at the solution. Is it not amazing how a mole in England looks like a mole in Australia? Across the vast vista of biological potential, how did they "find" the same solution?

All life on Earth, all life across the universe, conforms to the law $P = F/A$. Life has no choice in this matter. Whenever this law plays a role in biology, whether it be our burrowing moles, a worm trying to push its way through some sodden soil, or a sand eel slithering silently through sandy sediment under the sea, all of them must observe the rule. Any biological solution that invokes this law by default observes it; life does not navigate to these solutions. The question, then, is whether the representation of this principle found in the biological solution allows the organism to survive long enough to reproduce. The mole that is born the shape of a sphere or with no front legs is not going to survive underground long. The sand eel that is too fat to burrow its way into sandy sediment will be gobbled by a larger fish and will also fail in the evolutionary struggle.

With the mole, for which $P = F/A$ happens to play a prominent role, there is only one broadly acceptable biological solution to the optimization of this equation: small animals with stubby, shovellike front feet. The animal did not navigate across a vast landscape of mole possibilities. It had to conform to the law, and the solutions that emerged through mutations from the first mole-like animal that dug underground optimized the application of this equation that was most effective at escaping predators, finding food underground, and crawling out of collapsed tunnels.

Variant moles with a genetic mutation that makes them more like spherical moles or fat, cuboid moles are less effective than their conspecifics are at burrowing. Ultimately, they are less competitive for territory and resources than those that apply $P = F/A$ more effectively. But those mutations can become effective if the environment changes. If our mutant moles find themselves in a hypothetical new environment where escaping predators by rolling down a hill, pangolin-style, is now the most effective way to escape, then a spherical mole, the circumference of whose rotund body is nicely described by $2\pi r$, may well find that its somewhat porky profile, which disadvantaged it in the burrowing struggle for life, now provides it with an advantage in the hill-rolling escapades demanded of it. Its circumference being described by the relationship $2\pi r$ is now the pertinent relationship that operates on the rolling mole. In this hypothetical way, organisms move from one set of conditions to another, across a chessboard of environments, observing a vast variety of rules; those mutants that either use these laws to their advantage or conform to them with greater effectiveness survive to reach breeding age.

Even with a narrow, straightforward set of physical laws, we can observe wonderful diversity. Increase the importance of $P = F/A$ to a burrowing animal whose ancestors had legs, and we end up with moleness. Do this to a legless invertebrate, and we end up with wormness. Earthworms are long and slender, but as limbless invertebrates, they move not with robust little legs like moles, but by using muscles that contract and expand, forcing their way through the ground. A juvenile worm can push more than five hundred times its body mass as it drives itself through the soil. Variety manifests itself in how laws operate on biological material with prior history, but convergence makes these different branches of life very similar. Even between different groups of life, vertebrates and invertebrates, moles and worms share long, slender, cylindrical shapes with pointy fronts.

Physics explains much about what we find in life, but can it also explain what we do not see? There is a conversation that some biologists have over coffee, or sometimes pints of beer. At first blush, the conversation may seem a little esoteric, maybe even somewhat bizarre. Yet the question they ask lies at the heart of everything that concerns this book: why don't animals like moles have wheels?

Take a look at everything around you that has to do with transport, and you can see that this question is not so strange. Cars, trains, bicycles, and wagons have them. Even planes, although they fly, land on wheels. These modes of transport, and many more, use these simple circular devices. So why has nature thoroughly rejected the wheel?

Wheels work tremendously well on roads, railway tracks, and flat surfaces, but most of the world is a haphazard mixture of hills, ditches, and other irregular obstacles. Among all these, the wheel comes up short. A wheel cannot get over any vertical obstacle higher than its radius; it just slams into the obstacle and stops—unless the wheel can be lifted over, much like how someone pushing a shopping cart will pull back on the handle to launch the front wheels over a curb. Without the complexity of shopping cart and curb shenanigans, this simple problem can be described in an equation in which the force F to push a wheel over an obstacle is given by:

$$F = \sqrt{(2rh - h^2)}\ mg/(r - h)$$

where h is the height of the obstacle, r is the radius of the wheel, m the mass of the wheel, and g the gravitational acceleration (on our planet, 9.8 meters per second per second [m/s^2]).

Make the height of the obstacle the same as the radius of the wheel, and this equation tells us that the force becomes infinite. You're stuck.

The landscape of our planet is full of irregularities, and there are more obstacles the smaller you are. A wheeled ant or ladybug would have a quite terrible time trying to get over grains of sand and soil particles, even if it were a four-wheel-drive creature.

Other problems confound our wheeled critter. Muddy soils, sand, and anything that offers resistance to the rolling of wheels will slow it down. A hungry, bounding four-legged fox would delight in a wheeled rabbit stuck in an English field, flinging mud all over the place as the rabbit wheels go round and round.

Legs offer that all-important possibility of essentially prodding your way across the landscape, maneuvering left and right in zigzags and reversing easily whether you are being chased by a predator or avoiding a muddy patch. In extreme situations, like the mountain goat perched high on a ledge a few centimeters wide, its legs deftly navigating the footholds irregularly distributed across the craggy surface, the advantages of legs over wheels are all too clear.

It is tempting to think that the lack of wheels in life may simply reflect the ancestry of all land animals. Perhaps land animals are merely constrained by the legacy of their ancestors, a limb joint that is simply unsuitable for making wheels in the first place. However, there does not seem to be a fundamental barrier. We could imagine an early fin that flips around in a rotating manner similar to a person swinging the arms in a circle as the first fish began sojourns onto land. Increasing efficiencies would be achieved by better performance in rotation and stronger limbs. Eventually, the structure would evolve into some sort of wheel that would evade the problem of nerves and blood vessels getting entangled into an awful mess as it rotated.

Evolution does experiment with wheel-like contraptions. Across all continents of the planet, excluding Antarctica, dung beetles gather up balls of animal dung into spheres and diligently push them across the landscape back to their burrows either to eat the dung or to use it as a brooding ball for their young. Navigating using the Milky Way, the navigation itself a stunning feat of the evolution of visual cues, the beetles can push an impressive ball, between about ten and a thousand times their body weight. These balls of dung show us that on flat, dry terrain, evolution does test rolling motions. Rolling works where the ground is predictably flat.

Similarly, wheeling across the deserts of the Earth are tumbleweeds, near-spherical parts of plants that break away from a mature plant and head off, driven by the wind, across the landscape to colonize new terrain. Most of this plant material is dead, and this arrangement has an advantage, since when the weed has stopped moving, the seeds can become detached from the dead material and drop to the soil to germinate. More than ten families of plants produce tumbleweeds, and all these plants live in arid and steppe-like regions of the world, where vistas of flat horizons provide the ideal environments for these vegetative emissaries to roll unimpeded across the land.

These examples, although not wheels in the strict sense, show that evolution, for some inexplicable reason, did not overlook the circle as a solution to moving something across a landscape. But evolution just did not go far with this form.

It would be tempting to follow up this discussion with another question: with all that irregular terrain, why don't animals just build roads or at least road-like flat thoroughfares? Richard Dawkins suggested in an interesting rejoinder that road building is just not selfish enough. If you build a road, someone else may appropriate your hard-won efforts, meaning the energy put into constructing it was worthless. Roads in our own society are very much a product of government action on behalf of everyone—you pay for the road even if you do not use it. With private roads, you must pay a fee to the builder, a transaction difficult for animals that have not invented economics. The argument is persuasive, but probably a much simpler answer is that for road building to evolve in animals, they must either have wheels to begin with or be evolving wheels to provide a selection pressure for road building. However, as the preceding paragraphs have shown, organisms do not even start to evolve wheels as appendages in the first place. There is no selection pressure for road building.

People who wonder why rabbits do not have wheels are usually the same type of people who wonder why fish don't have propellers.

This is a slightly surreal, but no less interesting question. After all, propellers are ubiquitous in our ships, boats, and even the smallest little paddleboat on a Greek summer holiday. Why doesn't fish evolution converge on propellers to get through water as moles converge on small, cylindrical shapes and paddle-like front feet to get through dirt?

Propellers are not very efficient. Rotate a propeller too fast, and the flow around it is broken up. When the water *cavitates*, or forms bubbles, they form around the tips of the propeller, reducing the thrust to drive a ship forward. On ships, propellers typically reach about 60 to 70 percent efficiency at best. Compare this with the way many fish swim: by flexing their body and inducing an undulatory wave to pass along it or part of it. This locomotion can be over 95 percent efficient. Contorting your body and writhing through the ocean turns out to be an efficient way to escape predators or reach food before your competitors.

Before we discard propellers as a seemingly unintuitive possibility—an image that even elicits ridicule when we imagine propellers stuck to a fish—there are examples of rotating structures that drive living things though liquids. Dangling from the sides or ends of some microbes, such as the much-studied species that lives in your gut, *Escherichia coli*, are flagella (in the singular, flagellum), whiplike appendages that spin at amazing speeds of about a hundred turns per second. They propel the microbe forward through fluid at up to six hundred microns per second or about two meters an hour. But don't be too unimpressed—seen in another way, that's about six hundred body lengths per second, which for someone of your size doing the same relative body-length-scale rate of movement equals roughly thirty kilometers an hour, which is a fast human running speed.

The flagella are an extraordinary product of evolution. Tiny motors embedded in proteins in the microbial cell, itself about a micron long, rotate the long protein units that make up the flagella. Some microbes have just one flagellum, some have many flagella, and some can grow them at will, depending on whether they want to settle down or propel themselves somewhere new, perhaps to escape a toxin or find food. By briefly reversing the rotation of the flagella, a microbe can cause itself to tumble, changing its direction in its quest for optimum conditions for growth.

Despite the superficial similarities between a ship's propeller and a microbial flagellum, which might give us pause to wonder why fish didn't use the same trick to get around, there are vast differences between the environment that a microbe inhabits and that of a fish.

Imagine you were told to swim the length of an Olympic-sized pool, but not any pool—one filled with molasses. As you climb into the gooey pit, you allow yourself to sink in. Then as your whole body is submerged, with a mighty effort, you push your arms behind you in the first stroke. You move forward, but just a few paltry centimeters. As you pull your arms to the front through the syrup to begin the second stroke, you find that your arms' forward motion pushes you back again, by the same distance. As you continue your efforts, you remain fixed, oscillating back and forth in insignificant movements as your arms torturously push forward and backward against the gloopy substance.

At the scale of a microbe swimming in water, this is what everyday life is like. Water behaves like a viscous liquid. A conventional ship propeller at this scale would not work, because it relies on pushing water backward, virtually impossible in a syrup-like liquid. So the comparison between the propeller and the flagellum is misleading. The flagellum should instead be thought of as more of a way of corkscrewing a microbe through the water rather than as behaving like a propeller to push something forward by imparting momentum to the water behind.

The efficiency of the flagellum is poor, about 1 percent, much less than a typical propeller on a ship. In the microbes, we do not see nature using propellers; instead, we see a device for making something move at small scales, at the syrupy scales of what physicists call *low Reynolds numbers*. A flagellum is a particular solution for little things that want to move, but it nevertheless shows that low-efficiency rotating structures for propelling things through fluids are not unknown in nature. At larger scales, at high Reynolds numbers, when water viscosity becomes less important and the fluid behaves more like the water in a swimming pool, propellers work, but they still remain less efficient than slithering through the water like a fish.

Wheels and propellers get us into some trouble though. Although we can imagine how a wheeled rabbit or propeller-driven fish might evolve, we still cannot conclusively tell whether the absence of these creatures is caused by a development barrier, an insufficient versatility in the tool kit of life to make the leap to these contraptions. It is much easier to experiment with, and trace the evolutionary path of, something that does exist. However, we find compelling physical reasons to suspect that even if the developmental path to wheels and propellers were available, life would eschew these devices for legs on land and body waves in water.

The recognition that simple mathematical or physical principles might determine what organisms do and don't look like has occurred to scientists before. In 1917, a brilliant Scottish mathematician, D'Arcy Wentworth Thompson, published a controversial book titled *On Growth and Form*. In its prolific and fascinating pages, he demonstrates the myriad ways mathematical relationships and scaling can be found in life. He considers the regular ("equiangular") spirals that define the shapes of shells, from snails to the extinct ammonites. He explores the shapes of horns and teeth and the mathematical relationships in the growth of plants. He even draws fish on grids and, by shearing the grid in different directions, he shows that the now obliquely shaped novelty resembles many species found in the natural world. The point of his book is simple—that biology can be described by mathematics and that all living things conform to simple patterns of scaling and interrelationships between their different dimensions. When his book was published, it was rather radical in an age transfixed by Darwinism. Even today, people are somewhat unsure what to make of it.

Thompson's book is underpinned by the same logic that runs through much of this book. Physical laws, incontestable and unavoidable, operate in life. Organisms that transport themselves on legs must scale up in a certain way to counter the laws of gravity, as must trees. Hydrodynamics dictates the shape of a fish. A regular spiral in a snail will always conform to some sort of self-similar pattern along its length. Not surprisingly, if we go about measuring the shapes and forms of organisms, we find some repetitive and similar features. But there are, however, two things that Thompson does not do. He does not really try to articulate *why* his observations apply to life, and apart from his intriguing graphical shearing and rotating transformations, he does not suggest a realistic mechanism of *how* they might have come about.

He does, throughout his book, explain that many of these relationships are caused by the environments in which things live. For example, he says that plant growth is a response to movement against gravity and to the forces experienced during growth. Although he never forcefully drives home the point that these relationships must emerge from the tight coupling of evolutionary processes and physical principles, his book provides the first serious attempt to demonstrate the mathematical regularity in life.

The second question, on *how* all this happens, he leaves alone entirely, and this has caused some trouble. Why he never visited this question is uncertain, but we might equally ask why he was obliged to. His interest was in demonstrating how evolution leads to symmetry and predictable form. That was probably enough for one book. In the pre-DNA age of the 1910s, to have speculated on how these shapes came about would have taken the discourse into the territory of phantasm.

Since then, however, his critics have suggested that this lacuna implies that he implicitly rejects natural selection. His mathematics-infused assessment of life apparently leaves no room for the process of Darwinian evolution and its exploration of wondrous forms. Some observers have even claimed that his book supports *vitalism*, the belief that life has some sort of force or substance, discrete from non-living matter, as the component that animates it. Inanimate objects do not mysteriously transmutate into others, a rock into a metal bar, for example. So if Thompson believes that life is just mathematical relationships bundled into organic form and he does not invoke natural selection, he must be saying that there is something else in life, some vital force that allows for one form of life to transform into a different one.

These critiques probably miss the point. Thompson's analysis says nothing of the sort. He merely points out that life, however it works, whatever process is at the heart of evolution, results in forms with predictable shapes. He observes that life is not boundless, but tightly hemmed in by laws. These rules account for the conspicuous symmetry and patterns we find in life. They explain instances of convergent evolution, and they might also help explain the lack of certain features in the biosphere. Nevertheless, with modern molecular and genetic insights in biology, it is worth asking how these things might come about. With our new insights, we are much better placed to understand how physics can shape life.

How does evolution achieve these feats of transformation? About a century ago, if you had asked this question, you would have been met by a standard Darwinian response. The information in life mutates, maybe by errors in reading the code or by some environmental assault, such as a damaging chemical. These small variations produce different offspring, some of which are better adapted to the environment than others. Those that are better adapted survive and reproduce, and those that are maladapted die. This incremental process of selection fashions life into new and exotic forms.

This Darwinian synthesis, despite refinement and improvement in our understanding of its mechanism, remains fundamentally accurate. However, our understanding of the mechanisms has been augmented in considerable detail during the last few decades by a growing knowledge of the process of animal and plant development, called *evolutionary developmental biology* or, sometimes, *evo-devo*. For some reason, I find this fashionable abbreviation dreadful.

By studying embryos and how genes switch on and off during development, biologists have shown that life is built up of simple modules, basic construction bricks that can be modified into a range of forms. The paradigmatic shift at the heart of this work was the realization that the code of life is not just a giant length of DNA read from one end to another with millions of bits produced and then self-assemble into life. Darwinism does not need changes in single units of this genetic code over vast tracts of time, an accumulation of small alterations that may seem improbable. Instead, genes are switched on and off, controlled by a whole battery of regulatory genes that can turn two very similar pieces of DNA into two ultimately different forms.

The difference between you and a chimpanzee does not merely reside in the 4 percent difference in the DNA code between the both of you, but resides in how sections of the other 96 percent are read.

Evolutionary developmental biology has revealed much more. Not only can life do clever things in the way it reads different parts of the same DNA code, but scientists also discovered that the instruction manual of life is structured into a grand hierarchy. Some genes cause differences that are small, but some genes control the whole pattern of development. None are as dramatic and impressive as the Homeobox (*Hox*) genes. Found ubiquitously in animals, from houseflies to humans, these *Hox* genes control limb development—the legs in a fly, the fins in a fish, and your hands and feet.

The extraordinary ability to modify entire biological architectures with some simple gene changes goes some way to accounting for the versatility in many important pieces of a living thing, and one of the most self-evident in animals is the limbs. In his wonderful book about this field, *Endless Form Most Beautiful*, Sean Carroll explores how appendages are merely modifications of the number and shape of fixed modules (digits) on a common design. Produce three long and spindly digits, and you have the foot of a crane, but put them at the end of a wing, and they now anchor the airfoil of a swooping falcon. Bury them in a flat flipper, and a sea turtle can now cross an ocean; reduce the digits to two and put them in a hard casing, and you have the foot of a camel striding across the baked surface of the Arabian Desert. These developments come about by subtle modifications of the *Hox* genes; the number and timing of the expression of these genes modify the limbs that an animal ends up with.

Evolutionary developmental biology's contribution has been to link development with evolution, but it also provides the beginnings of a better explanation for how evolution works. Like a Transformer in a Hollywood movie, by building life from repeating modules—whether they be digits, vertebrae, or even the cells that make certain tissues—changes in these modules can effect radical reorganizations as organisms exploit and move into different environments. Whether the life is in water, land, or air, these modules can be reconfigured by mutation to generate the architecture of living things that can persist in places that may be poles apart.

But what has all this to do with physics?

In evolutionary developmental biology, we have the groundwork for harmony between evolution and the laws of physics. Consider our mole. To claim that the laws of physics drive moles in Europe and Australia to the same solution to the problem of optimizing $P=F/A$ is to invite the question "But how?" How can a mole whose relatives are shrews and weasels look so starkly similar to a mole whose relatives are kangaroos and koalas? Isn't that uncanny? And how could this resemblance occur with incremental changes? Once an animal is locked into looking like a shrew or a kangaroo, is there no turning back? How could two lineages bring about the changes needed to converge on a similar form?

As the laws of physics work relentlessly on new forms that emerge from the old, the modular design of life makes possible transformations that do not require entirely new designs. The existing tool kit of limb units allows for front legs that are strengthened and wider, producing an animal design that is a tunneling mole with no alteration in the underlying basic units of life. Some of this variation may be possible merely by changing the expression of particular genes with no radical mutations, but the evolutionary study of development also shows us that there is the possibility for these more cardinal changes. Evolutionary developmental biology goes further, suggesting that organisms may not be so tightly bound by their prior development as was once assumed and that modular rearrangements can help circumvent the constraints of genetic heritage. Such flexibility might increase the opportunity that physics has to creep in and determine what evolution can and cannot do. Lacunae in the evolutionary experiment may not always be explained away by a simple developmental barrier, but instead might sometimes be genuinely less adaptive from the standpoint of physical principles.

Within the evolutionary story of whole organisms, there are other mysteries about the shape-shifting behavior of life. Prominent among them is how an organism in one environment that has a certain set of dominant laws can diverge from its ancestors and make the transition into another, where a different set of rules comes to the fore without the animal ending up as a mushy evolutionary mess of jerry-built chaos?

One transition is impressive by any standards—the invasion of land. This transition provides a particularly lucid example of a decided shift between two environments, water and continents, with different physical requirements. The science of evolutionary developmental biology provides a clearer picture of how organisms can make these strides from one set of physical conditions to another.

Moving from a watery habitat to a new home on land presents life with some very profound changes in the physics of how one exists. There is room for intermediate transition stages. Muddy ponds and tidal regions on land offer something of a twilight zone of a water and land existence. In the water, some fish walk along the sea floor looking more like land animals. But the complete transition from water to land is nevertheless profound.

As an animal moves to land, a noticeable difference is the enormous gravitational field that is no longer counteracted by its buoyancy in water. A full 9.8 m/s² pulls down on an organism as it seeks to wrench its body from the water and make progress over a muddy foreshore, experiencing what astronauts feel, newly returned from a space station and needing to recline in a seat for a while lest they stumble under their own laborious weight. If this value for gravity still seems a little arcane, do 250 push-ups right now with one arm. Yes, that's right—that's the effort of pushing up against 9.8 m/s².

In the water, the overall forces that acted on the fish before it launched an expedition onto land were given by the expression:

$$F = mg - \rho Vg$$

The term mass multiplied by gravity (mg) expresses the weight of the organism pressing downward. Then there is the buoyancy of the fish, the upward force that offsets the downward force of its weight to allow it, with minimal effort, to glide through the ocean with a flip of its fins. That buoyancy is the term ρVg, made up of the density of the fluid (ρ) multiplied by the volume of water the fish displaced (V) and the gravitational acceleration, or g. You will have experienced how buoyancy relieves you somewhat of gravity as you wallow and float in a swimming pool or the ocean on a summer vacation. Now on land, ρVg has all but vanished; the buoyancy of an animal in the relatively thin atmosphere is irrelevant. All that is left is the downward force of gravity, mg, the weight of the poor, beleaguered creature. The animal must make up for losing ρVg by the sheer effort of pushing itself off the ground.

Here we have a transition as difficult as any, between two places that impose a different concatenation of laws. Now we have a conundrum. How can the physics of these two worlds and this divergence in evolutionary trajectory be reconciled? How can this transition occur?

The answer may lie in small mutations, or it may instead reside in dramatic changes. Evolutionary developmental biology has shown us that the modular design of life, in which whole segments of an organism, such as its limbs, may be modified using alterations to one or a few genetic units, is well suited to allow these transitions between environments that seem very dissimilar.

In stunning experiments, researchers have explored how fish may have made the transition to early four-legged animals, the tetrapods. The *Hox* genes that control limb development are controlled by a set of regulatory genes that tell them when to switch on and off. These global control regions harbor strands of DNA, which mastermind the production of hormones that themselves influence the development of appendages in the developing embryo. The scientists found that particular genes, such as the *CsB* gene, could be transplanted from zebra fish into a mouse and could drive the development of limbs. They found that the gene oversaw further genetic expression in the autopod, the part of the limb that produces the digit and the essential parts of legs for walking on land. In inverse experiments, enhancer genes from mice can be transplanted into fish and will drive development of the fins.

These impressive, if ghoulish, experiments show us that the genes that regulate fins and limbs are similar and very ancient. Deep in the history of evolution, the basic genes that evolved to control the general architecture of developing appendages in animals continue to regulate limb development, whether that is a fin or a leg.

Now it would be apropos for the reader to exclaim surprise. How did an early animal know that the genes that control a fin would eventually be required to do the work of making limbs to walk on land? How is versatility in a genetic code locked into early animals that would one day be used to take on the challenge of very different physical environments from the ones in which they emerged? This so-called inherency, as it has sometimes been called—the curious apparent foresight that evolution seems to have for challenges yet to come—might seduce anyone into accidentally thinking there are grander minds at work.

Remarkable though the feat of turning fins into limbs may seem, perhaps we should not be too amazed. We are not asking life to make the move from the oceans to living on the surface of a neutron star with gravitational forces many millions of times greater. No radical physics is being asked for here. Provided that the environment is not inimical to the modules from which a multicellular organism is made (not too hot, cold, acidic, and so on), then the basic chassis from which animals are constructed can be modified to suit the laws prevailing in many environments. Those genes that evolved to make a fin are now being asked to regulate in such a way as to make the bones separate into distinct digits and fashion them thicker to withstand a world in which the buoyancy term, ρVg, in the force equation is now irrelevant. That transition is difficult but not utterly alien. We are removing one term in an equation in which the weight of an organism (mg) becomes more pronounced, but it was nevertheless there from the beginning of evolution.

As the first fish pulls itself onto land, those offspring with stronger bones and muscles will more effectively yank themselves off the ground and drag themselves along to find food and maybe some shelter from the midday sun.

How fish made the transition between swimming and walking is still somewhat of a matter of debate. Intermediate forms of locomotion are seen in some mudskippers and tidal-flat-dwelling animals even today, including tail-flip jumping, an awkward way a fish can hurl itself into the air on its side by a good, sharp whack of its tail against the ground. The slightly more graceful, but primitive prone jumping is similar to tail-flipping, but here the fish uses its tail to throw itself upward and forward with its belly facing down so that it is right side up. It can at least see where it is going, and it does not land on its eyes. By contrast, some fish, such as the pink frogmouth, *Chaunax pictus*, walk across the sea floor very much like land animals do, a feat suggesting that fish learned to walk even before they emerged onto land.

Ultimately, the fundamental physical alteration that was needed was the transition between fins and limbs that were sufficiently engineered to hold an animal up against gravity with the requisite flexibility to allow walking. Once that organism can move around on land with primitive limbs, alterations in animal size from this point on are a simple scaling between the mass of the organism that defines the weight, mg, it must hold up and the thickness of the bones and the strength of the muscles to hold them up. These are the scaling laws that Thompson so vividly observed.

Nonetheless, in the history of animal life, among the many independent vertebrate invasions of land, only one lineage led to us. In this evolutionary detail, we have evidence that the transition between water and land is not trivial. There is no foregone inherent capacity in the early evolution of marine life to move onto land. The laws that operate in both environments are sufficiently different and the selection pressures so enormous that it seems this transition required several experiments to succeed.

The notion that physical limits could prevent animals from making some transitions is not wild conjecture. Even on Earth, there are environments, from the boiling volcanic pools of Rotorua in New Zealand to the acidic rivers of the Rio Tinto in Spain, where animals cannot live. In these places, only microbes reign supreme. Places like these demonstrate that physical limits can be too extreme for animal evolution. Evolutionary inherency is impressive, but not limitless.

Dealing with the physics of a disappearing buoyancy term in the equation that defines one's movement on land is not the only challenge in clambering out of water. Apart from the difficulty of walking around, there are other problems, too. Under the glaring sun, water evaporates without mercy from the scaly surface of our new land denizen. The latent heat of vaporization, which is the energy you need to turn water from a liquid into a vapor, is high, 2257 kilojoules per kilogram, about ten times higher than for many other fluids, but nevertheless the sun's energy is more than enough to drive off water, as you may have experienced drying off in the summer sun after a swim. Without drinking water, dehydration follows. A paradox for a creature that has chosen to leave the security of bountiful water is that it must now remain close to abundant water to prevent itself from desiccating. And the physics doesn't end there. As water evaporates from the surface, it also carries off energy, thus cooling the animal. It had better beware, lest it get so cold that it cannot move.

In the blaze of the solar furnace, there is a balance to be had. The creature must use the sun to warm itself, but the loss of water threatens dehydration. It must evolve a thick impermeable skin that delays the loss of water. These innovations can be made good by changes in the genetic modules that control the development of skin cells. All these weird, new experiences now confront our nascent tetrapod as it seeks to navigate the nooks and crannies so unfamiliar against the predictable three-dimensional regularity of its former aquatic realm.

Once a fuzzy blob in the aquatic void, the animal that now stares bemused at the star in the sky is bathed in UV rays, short wavelengths of light previously screened out by murky coastal water. It stares into the cold, uncompromising eyes of a relentless equation:

$$E = hc/\lambda$$

which describes the energy of light. This demon has become part of its world more than ever before. The energy of light (E), given by multiplying Planck's constant (h) by the speed of light (c) and dividing by its wavelength (λ), determines that light with a smaller wavelength, such as UV radiation, has much more energy than does light of longer wavelengths. Those wavelengths now impart their energy to the surface of the fish, and with these rays comes the chance of radiation damage, familiar to you and me as sunburn, even cancer.

The damaging effects of radiation are a strict physical rule. Shorter wavelengths of light have higher energy and can do more damage to molecules than can longer-wavelength light. This line of logic applies as much for our landlubber as it does for any other organism across the universe. In emerging onto land, where there is higher UV radiation, the animal might mutate to produce more pigment to protect itself. We do not know what the pigmentation of the earliest land dwellers looked like, but chemicals such as melanin might have been used. This pigment, found in our own skin, helps prevent radiation damage in life forms as diverse as fungi and animals. Its dark color causes suntans and the natural dark coloration of people who inhabit places where the rays of the sun are more unyielding, such as across Africa and Asia. Its chemical structure, a complex network of carbon rings and chains, probably evolved early on from the oversynthesis of an amino acid such as tyrosine, a pathway as ancient as the formation of proteins themselves. Even here, we find evidence of changes in existing genetic structures, the commandeering of old biochemical pathways for new roles and challenges in novel environments, but ones that are not so dramatically different that the laws of physics overwhelm.

No matter where the UV-screening compounds came from or what their chemical structure or their color is, they are all driven in their evolution by the simple relationship that inheres in $E = hc/\lambda$. Not surprisingly, regardless of their origins, they all share some common attributes. Most have long chains or ring structures of carbon atoms that absorb in the UV radiation range, chemical structures mandated by the carbon bonds that most effectively absorb UV radiation, which are those with delocalized electron systems. Here we see evolutionary convergence at the chemical level, ultimately fashioned by the exposure of large animals to radiation in the environment.

Clever adaptations to live on land did not have to be created from scratch. UV-screening compounds came from familiar biochemistry. Many marine organisms are exposed to some UV radiation. The clear open oceans allow UV radiation to penetrate far into the water, and except for the deepest-living ocean life, many marine organisms are exposed to some level of UV radiation. An assortment of the physical laws that operate on land exist also in the oceans; it is often just their magnitude or some components of the laws that are modified.

Even problems that seem original may not be outlandish. Skin that prevents desiccation seems a distinctly terrestrial innovation, but of course fish do not want their innards diffusing into the sea. There is already a selection pressure for a robust barrier between the insides of fish and the outside world, a barrier that can be augmented in the dry environment of land.

Once life had emerged onto land, its modular capacities to adapt continued to provide the promise for evolving in exciting ways. In experiments that tracked the expression of genes in python embryos, scientists at the University of Florida showed that the *Hox* genes to make limbs are still found in the DNA of these snakes. By naturally suppressing the production of a limb enhancer called *Sonic hedgehog* (*SHH*) (yes, biologists do have a strange sense of convention for naming genes), pythons prevent the limbs from forming. Through this rather elementary modification of gene expression, we move from four-legged animals to snakes and, with it, the ability to slither, unimpeded by limbs, through undergrowth, up into trees, and through sands and soils. The latent *Hox* genes may even explain the discovery of fossil snakes with limbs. Now long extinct, these ancient beasts probably regained their legs by merely reexpressing capacities that remain dormant in snakes to this day.

Almost certainly, the secrets of that other extraordinary transition—from land back into water, a lifestyle change that needed limbs to be reformed into fins as the buoyancy term, ρVg, became a fact of life once more—are partly to be found in the *Hox* genes of whales.

Evolutionary developmental biology has opened our eyes to how life can move from one environment to another, each containing a set of laws that in combination require modifications to the structure of life, achieved through a rearrangement of a few basic designs. They allow for the successful invasion of new habitats, all of which, though different in physical character, are still in gross form not too diverged from one another, sharing as they do the same gravity, atmosphere, and oceans that characterize our small world.

Charles Darwin concluded his book *On the Origin of Species by Means of Natural Selection, or the Preservation of Favoured Races in the Struggle for Life* (it does the book an injustice not to quote its full Victorian melodramatic title) with this observation:

> There is grandeur in this view of life, with its several powers, having been originally breathed into a few forms or into one; and that, whilst this planet has gone cycling on according to the fixed law of gravity, from so simple a beginning endless forms most beautiful and most wonderful have been, and are being, evolved.

Here Darwin draws us to two important inferences. First is the overt notion that the Earth began with a law, the law of gravity, and then somehow broke away from this simple beginning into something more complex. The separation was not deliberate maybe, but a lasting legacy that implicitly separates physics from biology.

His second statement, that from these simple beginnings "endless forms" emerged, is yet another implicit divergence from physics. This statement is easy to nitpick. Darwin was taken to literary flourishes every now and then; he was a good writer. And in some ways, Darwin was right. In detail, there are potentially endless forms. The possible arrangement of colors on a butterfly wing is probably limitless if we consider that every scale might have a different shade, hue, or color and be arranged in ever-so-slightly alternative tapestries. Like human faces, there is endless detail.

But where he and I part is the general tenor of his conclusion. That the world emerged from something as banal and simple as the law of gravity and then exploded into an efflorescence of endless forms of biology is artistically captivating, but scientifically distracting. It is ironic that he picked gravity as his basic physical law, a law that plays an enormous role in shaping life at the large scale, from the scaling of animal sizes to the structure of trees. Gravity has indomitably followed the evolution of life from its very beginnings to the present day and made its unyielding mark on the form of things; it is a law that dominated the change in the character of life as creatures crawled from the sea to take ownership of the land; a law that has ensured that life on Earth is bounded and not endless.

Physical rules continue to shape the form of living things, endless in detail, restricted in form.

CHAPTER 5

BUNDLES OF LIFE

THERE ARE SOME NUMBERS so large we have no real way to comprehend them. If I tell you there are three poodles yapping outside my house in Bruntsfield, Edinburgh, you can instantly envision those creatures, bubbly haired and boisterous, tottering and sniffing around on the cobbles. However, if I tell you there are about 3.7 trillion cells in your body, now the number is a just a fuzz, a number with no meaning. It is just gargantuan, so enormous that you cannot imagine this collection of objects.

This is the realm we enter when we ask what living things are made of and from what units they are assembled. From the individual ladybugs, moles, and ants, we enter into the cellular dominion, the scale of life that is, as bricks are to a house, the next level down at which we find order in the makeup of life. At this level, too, we see that what was once a landscape of great complexity has given up its secrets to more easily understood physical principles. Through their manifestation in equations, the cellular world has become amenable to prediction, and the role of contingency has faded.

We have no knowledge what Robert Hooke was expecting to see, if anything, when he took his microscope to some dried-up cork in the 1660s. Peering down the lens, he saw serried, or crowded, ranks of holes, rows and rows in regular formation, like little rooms in a monastery. Indeed, that was precisely the thought that struck him, for he called these minuscule compartments he saw before him *cells*, from the Latin word *cella*, or small room. Hooke had no inkling of the significance of these structures, and although he drew, alongside his famous fleas, diagrams of cells in his book *Micrographia*, published in 1665, no one else could see the significance of them, either.

He was not alone in observing microscopic forms. Across the North Sea, another inquisitive scientist, Antonie van Leeuwenhoek, a Dutch tradesman, had fabricated pocket-sized microscopes made from glass beads. With this tool, he had sated his interest in this hidden universe by turning his new contraption to a multitude of things, from pond water to scrapings from his teeth. What he found in this previously unseen microcosm were creatures, "animalcules," or animals in miniature, many of which seemed to move. He even observed how he could kill them: by exposing them to vinegar, he could arrest their movements. In a series of letters to the Royal Society, he documented his findings. As was the case with Hooke's discoveries, it required creative thought to understand the significance of van Leeuwenhoek's observations. To many, these dwarf living things were evidence of a miniature universe of animals for sure, but beyond that passing fancy and fascination, there was little more to be said.

It would be another two centuries before the importance of this cellular cosmos was recognized—before the world would understand that those little holes seen by Hooke and the diminutive flitting animals documented by van Leeuwenhoek were the manifestation of the same phenomenon: tiny bundles of life, the unit of organization from which creatures are made. Hooke had seen the outlines of dried, empty shapes in his cork, but van Leeuwenhoek had seen independent living creatures. The Dutchman's observations were of special significance since he had observed microbes, which through the work of Robert Koch, Louis Pasteur, and many others would turn out to be the harbingers of disease and the microscopic factories from which beer and wine spring forth.

Observations of much more of the living world would bring this microscopic realm into front view. In 1839, Theodor Schwann and Matthias Jakob Schleiden, two German scientists, proposed the cell theory, an elegant idea that strung together several related observations that until then had remained disparate. Their idea, radical at the time, was that all living things comprise at least one cell or more; that cells are the fundamental unit of structure in all life and the source of all the varied functions in organisms; and that cells come from pre-existing cells by some method of multiplication. Because the mechanisms behind all these conjectures were yet to be demonstrated, the theory was radical. Today, its tenets are so banal as to be obvious. We would recoil in describing these observations as the cell theory, but we would merely observe the fact that life is made from cells. The theory is just an accepted fact of biology, an observation that underpins the whole field of cell biology.

With the benefit of another 150 years or so of research, we now know the principles that govern this most fundamental unit of life. We understand the physics and biology that run through it and where in this picture the contingent historical quirks of evolution could have their say.

But first, we might ask why. What caused this strange packaging of life into bundles that enthralled Hooke, van Leeuwenhoek, and all those who followed them in this journey to a miniature world? One simple cause is the basic physical principle of dilution. Pour some bubble bath into a tub of water, and soon its color is all but extinguished as its molecules mix and disperse into the water. So too on early Earth, molecules would generally have been diluted into the oceans, rivers, and streams. These molecules would need very special locations, such as the inside of rocks, to be packed close enough to react with other molecules to build more-complex chemical machines. If those early replicators ended up inside a small container, then not only would they stay concentrated, but now the world would be their oyster. These cages of molecules could move out into otherwise more foreboding, diluting environments, like the oceans.

The cell was, in simple terms, the answer to dilution, an innovation that allowed for expansion in a world with abundant water in which nature's tendency is to dissipate. The universality of compartmentalization in this way is suggested as a defining feature of life and, with reproduction and evolution, the basis of all living things.

Cells are certainly not the whole story of the biosphere. On Earth, we find biological entities without a cellular structure. Viruses, small pieces of infectious nucleic acids bound up in a protein coat, go about their business causing diseases such as the common cold in people and even causing mayhem in microbes. Yet these small hoodlums of the biosphere need a watery cell in which to propagate, a host in which to multiply. Because viruses are unable to reproduce on their own, without cellular life, some people question whether they should be allowed into the exclusive club of life or should be relegated to the less generous name of *particles* or *entities*. Prions, incorrectly folded proteins that propagate the errant folding of other proteins in a sort of chain reaction, leading to such grotesque infections as mad cow disease, are also thought by some to be outside the scope of life. Like viruses, they wreak their havoc in the cellular domain, but are not cellular themselves. Without cells, they are nothing, mere misfolded proteins in the wind. It seems inescapable that at least in our world, cellularity is a feature of life.

We can easily comprehend the centrality of the cell in the phenomenon of life through a simple thought experiment. Imagine a hypothetical garden pond full of organic debris and material. Because of some strange weather, the material that blew into it, and some other factors, it develops metabolism. Raw materials get broken down to release energy. Even more amazingly, inside this pond, nucleic acids (DNA) evolve into a replicating system of information. Here, in someone's inauspicious backyard, a prototype cell has come into being! Regardless of the potential of this garden pond, it is not going anywhere, trapped in its earthly burrow. There is no chance of replication or movement to new energy resources and nutrients. The "cell" is shackled.

Our frustrated garden pond organism is a metaphor for any complex biochemistry confined to a physical space, whether that be a pore in a rock on the beach or an inscrutable hole in a hydrothermal vent. Any biochemistry that was randomly encapsulated within an early cell was released; it was set free to begin a planetary-scale expansion. It didn't *want* to do this, but when this event occurred, replicating molecules were now likely to become abundant and exposed to many environments across the planet. These environments would act on the molecules to drive yet more variations and more life. Cellularity not only provided a concentrating mechanism, but also provided the means by which evolutionary selection in its myriad forms could occur. In that sense, cellularity and evolution are inextricably linked.

An enticing question to ask is how the first cells might have come about. What formed that little cage in which molecules could congregate to make a self-replicating machine? Was it a particular contingent event in the history of life, a chance occurrence, or something more physically inevitable? To answer that question, we need to know something about how that capsule is formed and what it is made from.

Look down at the edge of a cell and around each of them. In every self-replicating life form on the planet, you will find a membrane, essentially the bag that keeps everything in. This membrane is no mere sheet of boring bound-together chemicals, a sort of shopping bag in miniature. The molecules making up the membrane are dramatic in their character and beautiful in their simplicity.

Within the membrane are molecules with a head and a tail, two distinct parts that are the secret of its clever chemical capabilities. The tail is a long chain of carbon atoms, one after another, in a string. These carbon atoms are hydrophobic, or insoluble in water. With no charge to make them soluble, just as oil does not mix with water, they will do all they can to avoid the substance. These tails are attached to the head, which has a different character. It is made with a charged group. In the phospholipids, an abundant class of these membrane molecules, the head is an atom of phosphorus bound to some oxygen atoms with a negative charge. The heads are hydrophilic; they like to dissolve in water So here we have a schizophrenic molecule, one end hydrophobic and one end hydrophilic. What is the molecule to do?

When it is added to water, something astonishing happens. The tails of the different molecules line up to face each other while the heads point outward into the water. A two-layered structure of these molecules spontaneously forms, in which the tails, now huddled together in the middle, can avoid the water, as is their wont, while the heads, pointing into the fluid, can satisfy their attraction to the substance.

This lipid membrane has not yet finished its tricks. It will not form an endless sheet, flapping aimlessly in the water. Just as a raindrop forms a sphere to minimize its energy, to minimize the surface-tension forces, this sheet of lipids will also curve in on itself and form a sphere. Spontaneously, and with no direction, the membrane becomes a sphere with a liquid interior. A cellular compartment has materialized. In the alignment of the lipids and their propensity to form a ball, the physical principles of ionic interactions between molecules and their tendency to minimize their energy drive these long molecular chains inevitably toward a cellular bag.

About four billion years ago, such spontaneously forming vesicles, microscopic spheres, would have encapsulated molecules that could replicate. Now ensconced in a chamber, the products of these molecular reactions would be concentrated. They would have accumulated inside the early cell, reaching concentrations that could allow a variety of reactions. Slowly, steadily, the metabolic complexity we observe in a modern-day cell emerged. The first molecules enclosed in these so-called protocells may have been active forms of ribonucleic acid (RNA), the sister molecule to DNA, a pioneering form of genetic code.

Once a cell had been born, evolution now did its work, not merely on lone pieces of genetic information floating around in a pond, but also on these early bundles of life. The cell itself now became the unit on which environmental conditions operated to drive evolution.

But, you might say, surely this is all speculation?

In the 1980s, David Deamer, at the University of California Santa Cruz, set out to find out whether simple chemicals, extracted from an ancient meteorite, could make membranes. Was the formation of a cellular structure—the answer to the physical problem of dilution—a chance event or something that was also physically inevitable? He chose meteorites, rocks that have blazed across the sky in the last leg of their journey from space and landed on Earth. Some are the most primitive materials in our Solar System. Among the family of these rocks are carbonaceous chondrites, black rocks known to contain carbon compounds. One such is the Murchison meteorite that fell in Victoria, Australia, in 1969. Long since a source of fascination for astrobiologists interested in finding out where the building blocks of life came from, these rocks, fabricated in the churning and swirling material of the early Solar System, were found to contain amino acids, the chemical units that make up proteins. But they also contain many more riches.

Deamer extracted some simple molecules similar to lipids. These carboxylic acids, like the lipids, have a charged head and a hydrophobic tail, but they are less intricate and generally shorter than the lipids used in modern cells. Deamer collected these molecules from the meteorite and added them to water. Spontaneously, the molecules gathered and formed vesicles.

Deamer had shown that membranes can spontaneously form, their shapes amenable to physical prediction and study. Not only that, but the molecules that make these cellular compartments were strewn through our Solar System in carbon-rich rocks. The ancient material fabricated from the swirling gases that orbited and condensed around the early Sun formed within them the very molecules to make cells.

Provided that the gases and other material in our own Solar System are not a freak accident, a one-in-a-million concoction, then we might expect the molecules of cellularity to form in any primordial cloud, ready to deliver their cargo of protocell material to the surface of any planet with a waiting abundance of liquid water.

Deamer's meteoritic haul of carbon compounds is not the only place to find these molecules. Experiments have shown a whole range of ways in which you can make them. Even on early Earth itself before life arose, chemical reactions may have produced membranes. Pyruvic acid, a simple organic molecule, will make membrane material when heated and pressurized. Apparently, these early cell ingredients were plentiful on the planet.

Although these ingenious and relatively straightforward experiments show how cell membranes can be formed, the exact place where this happened on the early Earth is a matter of contention.

Deep in the oceans, at the ridges that spread and split apart between tectonic plates, are hydrothermal vents. These chimneys of minerals are formed from fluids rich in dissolved metals and other chemical goodies that precipitate into towering mounds of rock as the fluid comes into contact with the cold ocean waters. Within these monoliths, searing water, still liquid at several hundred degrees Celsius because of the high pressures, gushes forth through pores and larger holes in the minerals, driving chemistry in these extreme conditions.

Some scientists think that within these sorts of environments, life's early chemistry began, taking advantage of the changing concentration of chemicals within the rocky pores. Eventually, when the first metabolic processes occurring on these mineral-rich surfaces were ready, they would have peeled off within the membranous coating to take flight into the wider world.

Some people eschew hydrothermal vents and contend that life began on the seashore, in the to-and-froing of tides washing in new molecules to concentrate in ephemeral pools or rocky outcrops. Others think that impact craters are a more likely place for life to begin, the intense heat as a space rock slams into the ground generating gradients of warmth and water circulation ideal for life. Perhaps Darwin's "warm little pond" literally manifested in a pool at the edge of some volcano on land was life's birthplace.

Whether the first self-replicating molecules and the cells that eventually enclosed them needed somewhere special to emerge or whether this process could have been happening in many places is a question to be resolved. But Deamer's experiments suggest that compartmentalization of chemistry into cellular structures is not a contingent result of a historical quirk, but is highly likely anywhere schizophrenic molecules with a combined love and hatred for water find themselves collected.

A mere bag of membranes and some enclosed molecules will not achieve much. As those cells emerged, in parallel with developing a genetic code, metabolic pathways to run those cells must have been forged. Within these cellular compartments, the pathways of life, the production lines of the building blocks of life, could become more complex, leading to the bewildering number of roads to the synthesis of the different components of life that characterize a typical cell. With our cellular membrane in place, we might now ask ourselves whether these earliest metabolic pathways were themselves strange chance events or whether there are predictable physical principles in these as well.

Take a cursory glance at a map of the cell's metabolic pathway on the internet, and all you will see is a vast number of arcs and lines linking the hundreds of products and intermediates of chemical roadways. These pathways produce everything, including amino acids that make proteins, complex sugars, the bricks of the genetic code, and various molecules that are the broken-down products of your food. Surely, these convoluted pathways are something different from their enclosing cell? In this labyrinthine mat of strands, surely contingency, more capricious historical events now have their chance?

Yet within this riot of complexity, there is astonishing minimalism. Some of the most ancient metabolic pathways found in cells, such as the reverse citric acid cycle, which produces the basic starting materials for many other pathways, use chemical compounds that could have been available on early Earth. The simplicity of these pathways, the chemicals involved in them, and the pathways' energetics suggest to many people that these reactions too could be universal. Their ubiquity across life appears to be more than just an accident fixed into the first cells.

Even greater store is given to these conclusions by the study of other parts of this patchwork of pathways, some of which also bear the imprint of universality. The glycolysis pathway, which breaks down the sugar glucose in the cell, and gluconeogenesis, the reverse pathway for making it, are ancient and very conserved routes for making and breaking sugars, essential for cellular construction and gathering energy. By testing thousands of alternative possible molecules and pathways, scientists at the University of Edinburgh showed that the routes used in life produce the highest flux of compounds of all possible alternatives.

These independent investigations show us that the metabolic transformations that emerged in the earliest cells on Earth were not mere flukes, but were the result of physical rules, although the information within them and the way they are structured may contain the imprint of their biological heritage. Nor do these sorts of conclusions support the idea that the pathways are accidents, fixed into living things early on and impervious to change afterward. Indeed, quite the contrary, it seems that they are quite flexible and that some pathways can turn into others with just a few mutations that alter the chemicals they use. If life wanted to escape the strictures of many of its existing routes to build itself, then it apparently could.

Instead, many metabolic processes in the cell are optimized in ways that have profound implications, suggesting that if life arose elsewhere in the universe, it too might land on the same, or very similar, networks and that we could a priori predict what those networks might look like.

For over three billion years before animal life, the microbes, these single-celled organisms with their attendant metabolic pathways, ruled the Earth alone. Yet on the outside, in their shapes, physical processes were slowly but ineluctably crafting them, just as they would eventually sculpt the moles and other complex life.

Talk to someone about a microbe, and the listener might, understandably, want to suppress a soporific yawn. Almost anyone unfamiliar with microbes thinks they are not much to look at. Yet within their microscopic domain, they show a farrago of fantastic shapes. Spheres, rods, spirals, filaments, and even bean-, star-, and square-shaped microbes abound in the world.

When the first cells moved out into the world from their birthplace, they too were shaped by laws, and their slavery to physics was a harbinger of how animals, although more complex, would later also be molded by inescapable rules. One first indefatigable effect of the environment on these new ambassadors of life was to keep them small. Most cells, whatever their shape, are minuscule. And they stay that way. But why?

There are many causes for a cell's diminutive size. A large bag is likely to collapse under gravity. The smaller the cell, the less likely gravitational forces will distort it and cause its contents to settle out. In this observation, we find gravity at play.

Cells have to contend with other challenges as well. They need to take in food and nutrients and expel any waste. Consider a spherical cell. The surface area is given by $4\pi r^2$, where r is the radius of cell. However, its volume is $4\pi r^3/3$, so as we increase the radius, the volume increases by a cube power compared with its surface area, which is squared. As a cell gets larger, the volume gets larger quicker than its surface area does. In other words, for every unit volume of its interior, there is less and less surface area to provide a means for nutrients to enter and waste to exit. The smaller you are, the more surface area you have in relation to your internal volume to make those vital exchanges. Added to the woes of getting larger is the problem of diffusion inside the cell. The larger you are, the longer it takes for nutrients to slowly migrate through the cell, perhaps from one end to another. So it pays to be small.

There may be other causes of smallness, but building a bag that does not collapse and ensuring the effective exchange of materials across the cell's boundary are two prominent benefits. Both of these results have simple physical principles behind them. It brings up the question of the size of the smallest possible cell. It cannot be so small that the DNA and other vital machinery cannot fit in. That smallest theoretical size is about two hundred to three hundred microns, just enough to fit in some genetic material and its attendant proteins and to allow for a few metabolic pathways. The estimate actually fits well with the smallest bacteria found in the wild, including *Pelagibacter ubique*, a miniature bacterium with a width of only 0.12 to 0.20 micrometers and a length of about 0.9 micrometers.

Being small counts, but it also brings some problems. What happens if you are running out of food and you want more surface area to mop up more of those nutrients lying around? One way is just to grow bigger, but as we have seen, the bigger a sphere you are, the bigger the problem with a declining surface area compared with your volume. The cell needs a way around this paradox, and it does this by not merely scaling up a ball, but instead by forming a rod-shaped body.

Think of a cylindrical microbe with a radius of one micron (μm) and a length of five microns. Now if I double that length to ten microns, the ratio of the surface area to the volume will reduce from 2.4 to 2.2, an 8.3 percent drop. There is slightly less surface area to serve any given amount of the volume. In contrast, consider a spherical microbe that starts with the same surface area as our cylindrical microbe that is five microns long and then expands in radius to have the same surface area as our elongated cylindrical microbe. Its surface-area-to-volume ratio has decreased from 1.73 to 1.28, a 26 percent drop. An expanding sphere has much less surface area to serve any given amount of its internal volume, and the surface-area-to-volume ratio drops faster than it does in a growing cylindrical microbe. If you want to increase your surface area, it is better to become a longer cylinder than a bloated sphere.

This is no mere physics fantasy. When we study microbes in the laboratory or in the wild, we often find that when microbes are starved, they become filamentous. Only some very basic math is needed to explain this shape-shifting behavior. Physics makes long microbes.

That is not to say that microbes cannot grow large and that they are always forced into long, spindly shapes. When we talk of smallness, it is all relative, and although in our world all microbes seem small, in their domain some get big. The giant bacterium *Epulopiscium fishelsoni* lives in the gut of the surgeonfish and grows to an incredible 0.6 millimeters across, large enough to be seen with the naked eye. Surely, this little beast has contradicted everything we have just said about the need to be as small as possible? However, a closer look shows that it is not defying the laws of physics. It lives in a gut with a high nutrient content from the fish's digested food, ensuring that even with the low surface-area-to-volume ratio, enough food gets into the cell. Throughout its membrane are folds, invaginations that hugely increase the effective surface area of the bacterium, a further modification that gets around the surface-area problem.

Some shapes that evolution has sculpted are surprising and arresting. The curvaceous shapes of certain bacteria, a mystery for many years, may allow them to attach to surfaces to form films, holding on in environments where water flows over them. The physical principles at work on these microbes emerge from the behavior of liquids and how they exert a shear stress (σ). Through high shear stress, liquids tend to rip the microbes away from the surface. These forces are expressed in equations such as the shear-stress equation:

$$\sigma = 6Q\mu/h^2w$$

where the shear stress (σ) is calculated according to the flow rate (Q), the viscosity of the fluid around the microbe (μ), the width of the channel (w) where the microbe is located, and the height of the channel (h).

Here, instead of nutrient needs, hydrodynamics comes face-to-face with life at the microscopic scale to shape and organize cells.

Across a planet with a legion of different environments, other physical properties may become just as prominent as, or more prominent than, fluid flow. Many bacteria find themselves in syrupy liquids, even more sticky than water normally is at their small scale. In a dried-out pond filled with organic gloop in your backyard or in the interior of an animal's gut, where the fluids are particularly viscous, there are many places where microbes are denied the free-flowing liquid in a river or stream. In these places, being a spiral shape seems to make moving around easier, as the microbes propel their way through their gluey world. Here too, a simple law, the behavior of particles in thick, viscous fluids, takes control.

Like moles and ladybugs, the single cells of the microbial domain are driven to convergent forms by simple physical laws, in some ways more easily distilled and discerned than those in more-complex multicellular creatures, but nevertheless just as uncompromising and predictable in the forms that they ultimately fashion.

We have overlooked one little detail in all these amazing shapes. Those membranous lipids from which these microbes are made are intrinsically flexible. They tend to form a sphere or squashy, amorphous blobs. Life had to stumble across one other critical invention. Surrounding most microbial membranes is a cell wall. Made of a substance called *peptidoglycan*, the wall is like a chicken wire mesh of linked-up sugars and amino acids. This wall provides the shape and rigidity to the cell and allows the cells to adopt these multifarious forms.

In the plethora of shapes we have toured, we find good explanations of how the cell wall emerged. The chicken wire coating may have given form to the first amorphous microbes. Unable to fix their shapes, they would have survived for sure, but they would have limped and bobbed aimlessly in their world. A cell that produced a substance that hardened its membrane could now be shaped by mutations and selection pressures to become spherical or filamentous or curved, enhancing its efficiency at attaching to surfaces, gathering food, or spiraling through goo. We can think of the cell wall as an adaptation that allowed life to be shaped by diverse physical principles, be they the laws of diffusion, hydrodynamics, or viscosity, to maximize its chances for survival and reproduction. For different microbial shapes, we can represent physical principles in equations and mathematically model how the principles influence the behavior and potential success of differently shaped microbes in their various environments.

As with ladybugs and other organisms at the large scale, physical principles are manifested within life, but through mutations, evolution also offers possibilities that allow the organism to reach reproductive age. Physical principles are as much a means to expand the repertoire of adaptations that enhance the chances for survival as they are an ineluctable condition of life's existence. These laws open up new space for invention. The cell wall opened up the vista of different shapes, each shape exploiting different principles that enhanced the chances of reproduction.

In this medley of microbes, do we see any chance events at work? If we reran evolution, we cannot say for certain whether all the shapes we see on the Earth would be explored. The giant *Epulopiscium fishelsoni* depends on being in an animal gut, and animals were not around until geologically recently. Like our ladybugs and moles, contingency allows for a range of details and modifications to be explored—a range that results in a lavish diversity in living things at the small scale. However, the range of forms is not endless, and those that exist seem to be channeled by some basic physical laws, which puts us in a position to make predictions. We can predict that on an alien world, cells would be small, and when starved, growing as a rod or filament might be one common way to circumvent the problem. If they are large, they will at least need to actively pump food in or make folds to maximize their surface area.

Where we might find contingency is in the structure of the membranes themselves. Since life first emerged, it has proliferated enormously and these munificent forms have explored numerous types of membranes.

A simple membrane surrounded by a cell wall is the choice of the Gram-positive bacteria, so named after Christian Gram, a Danish bacteriologist who first figured out a method of staining different bacteria according to differences in their membranes. The process makes bacteria easier to see in a microscope. Among the ranks of Gram-positive bacteria are the staphylococci, bacteria species of the genus *Staphylococcus*, a group of bacteria that can sometimes cause skin infections or food poisoning. They contrast sharply with the gram-negative bacteria, whose membranes are more complicated. Gram-negative bacteria have two membranes, with the cell wall between the two.

These two-membrane bacteria are everywhere and include the eponymous *Salmonella*. For many years, scientists have been intrigued by their strange double membrane—how did it get there? A possibility is that back in the days when bacteria dominated the Earth and were experimenting with their new planet, one bacterium got engulfed by another. By gobbling another bacterium without killing it, the host gained the benefit of the food that the bacterium might produce, maybe sugars or other products from its metabolism. The gobbled-up bacterium had the benefit of a safe home and maybe some other nutrients from its host. Think of a bacterium being engulfed. As the membrane from the host encircles it, it is now surrounded by two membranes: its own and the membrane of the hungry interloper. Once the prey is engulfed, we now have a double-membraned bacterium.

Neat though this story is, other people disagree and suggest that instead, the double membrane was a clever defense mechanism against antibiotics. These chemical products of microbial warfare are produced by many bacteria in the natural environment, and it is thought that they are used to fight for resources by killing or immobilizing their competitors. Because Gram-positive bacteria, with their less stringent single membrane, are generally more susceptible to antibiotics, the notion that the double membrane provided a more rigorous way of keeping antibiotics at bay seems to make sense.

In the membrane lipids themselves, the microbial world exhibits dazzling variety. In the archaea, a branch of the microbes that inhabit extreme environments, soils, and the oceans, the membrane lipids are chemically different from the bacterial membrane. In some archaea, the lipids are linked in the middle of the membrane, a position that is thought to give the membrane more robustness and integrity against high temperatures.

We could take long and discursive diversions to look at all the proteins found in cell membranes, the various sugars attached to them, and membranes' functions in grappling to surfaces and communicating with other cells. Not least, the membranes themselves are often covered in slime and meshes of sugar strings that provide yet more protection from the outside world. Sometimes, membranes spare bacteria from the fate of desiccation by trapping water, which is important for microbes that occupy rocks in hot deserts. Within and around a simple membrane, there is an efflorescence of biochemical ingenuity and diversity.

Looking on this wondrous profusion, we might reasonably think that once the chassis of a membrane is built—a wall that will enclose cell components, concentrate them, and provide a channel of communication to the outside world—then it is likely that chance can play a much greater role in all the membrane's details and accoutrements. Double membranes, links between lipids and slime layers—all these things and more can now blossom and experiment around that central cellular structure. Some of these developments might well be the result of random routes taken by evolution and less thoroughly channeled by physical requirements.

Even the cell wall may be a product of chance evolutionary changes. Does it have to be made with certain amino acids and sugars?

We do not know if its composition is a mere random event, established and carried on by evolution. The wall might be made of a selection of chemicals that simply happened to provide a reasonable barrier and a rigid casing. Once the job was done, it got fixed into life. Perhaps variations on its theme are also possible.

If we were told about a distant alien world with life, would we be confident to predict double-membrane life forms or the exact structure of cell walls? I suspect that eventually, once we know more about the biochemical origins of many of these adaptations, some may seem inevitable, likely innovations. Others may be contingent quirks of evolution with adaptive value in particular environments—quirks that, once discovered, did the job required of them. Rerun evolution on another world, and many appurtenances and jangling ornaments of cell membranes may be different in biochemical detail, but the cell membrane as a fundamental structure of life would be found in these other places too.

These many microbes dominated the planet for several billion years. Yet during this time, they did not cover the Earth with many isolated cells going about their individual business. One cell's waste could be another's food. Under the innumerable forces of evolutionary selection, from food scarcity to avoiding predation, these cells cooperated, forming multicellular aggregates. Nowhere is this spectacular division of labor better exemplified than in microbial mats, structures made up of layers of different microbes. These structures, thick gloopy mats in browns, oranges, and greens, often grow around the edges of volcanic pools, hanging on at the edges of boiling cauldrons. You may have seen them in the less exuberant form of green films growing on the sides of old buildings.

On the top of the films are usually the green photosynthetic microbes, trapping sunlight and using its energy to convert carbon dioxide into sugars. These tasty organic compounds find their way into the layers beneath, where other microbes, denied the light and warmth of the sun, carry out chemical transformations of the sugars in their dark underworld. In this way, waste and food are caught up in a cycle of give and take as each microbe has its place in this society of the small.

Cooperation, forced on the multitudes by the rigors of life on and within a planet, creates these cities of microbes. It is not uncommon for people to distinguish between microbes and the macroscopic world of animals and plants. These latter forms comprise "multicellular" structures. Yet microbes rarely live on their own and instead are often found cohabiting with other cells. Like the case with ants and birds, put microbes together, and complex, self-organized behavior emerges, patterns and movements that produce order that is more than the parts. These patterns and order arise particularly among microbes that can move and swarm across a surface in a coordinated way like a pack of microscopic wolves in search of food. Equations can be used to predict the coordinated behavior of the smallest cells. Here, in the cooperation of cells, we find mundane physical principles at play.

This cooperation is not mere chance, a fixed accident, but is an inevitable consequence of many cells inhabiting a planet. As they lock into a diversity of different nutrients and energy sources, it becomes a natural progression that the waste of one microbe might be used by another. An association is born. The sugars produced by a photosynthetic microbe will make it advantageous for a sugar-eating microbe to grow in close union with it. The biomass of this hungry little cell will become higher than it would become alone, in a sugar-poor pond. We need no superintendent to corral microbes together, to engineer their various collaborations and interactions. Place all this variety on a world, and cooperation emerges from the advantages this provides for each microbe. This self-organization can be modeled, predicted, and probed using equations. We would expect cooperation and aggregates—microbial mats, even—on any planet on which biological evolution has taken hold.

The key innovations of cells and their interactions leading to a world of cooperating microbes seem inevitable. So too their core metabolic capabilities. Those microbes at their cellular level show diversity in their membranes and their molecular clothing, but the cell as a discrete entity and the biochemistry that turned it into a self-replicating, metabolizing form is likely universal, forced by the laws of physics.

Before we close our tour of the principles underlying the cellularity of life, let us call into view one final question. This question is more complex, uncertain, and controversial but nevertheless explores an attribute of life that seems increasingly narrowed by physical principles. That question is how microbial life made the monumental transition to become those complex multicellular aggregates we call animals and plants. Was it inevitable and driven by the laws of physics?

Animals and plants are made of cells that are quite different from the cells of most microbes. The cells of the larger organisms are often lumped together as *eukaryotic* cells, their distinguishing feature being that their DNA is collected into a nucleus, a tiny subcompartment, or organelle, within the cell. They contrast to most of the cells we have been talking about so far, the *prokaryotes*, which lack a nucleus.

The domain of the eukaryotes, which, apart from animals and plants, include some single-celled living things such as algae, represents something of a revolution in the life of the cell. Eukaryotic cells tend to be much larger than the prokaryotes. Apart from their nucleus, the most noticeable difference between the eukaryotes and most prokaryotic cells is that eukaryotes have other organelles. Among them are the mitochondria, the power plants in most eukaryotic cells, including yours. Mitochondria make energy by burning organic compounds, such as the molecules from your lunchtime sandwich, in oxygen.

The eukaryotic cell is the child of a strange liaison in the early history of life. Engulfed perhaps by a feeding bacterium, the enclosed microbe would become the mitochondrion. This endosymbiosis was a curious internal agreement between the newly engulfed bacterium and its host, each giving something to the other. The engulfed bacterium gained the advantage of food and a homely environment, and the host got access to more efficient energy respiration using oxygen (aerobic respiration) provided by its new tenant. Many hundreds or thousands of mitochondria could generate energy in one cell, like a collection of power plants in a city, producing an energy revolution. This alliance released life from the restrictions of the energy-limited prokaryotes. Coupled with this, the growing genetic size and complexity of eukaryotic cells allowed for more-elaborate biochemical networks on which natural selection could act.

Thus, to make an animal, a concatenation of three extraordinary events had to occur. First, oxygen levels had to rise in the atmosphere to produce the gas that would unleash the energy-yielding capacities of aerobic respiration, the form of energy production used by you and me. Second, endosymbiosis had to happen. When a bacterium was engulfed, mitochondrial power stations could be made. They would multiply the capacity to make energy inside a single large cell, a feat needed to tap into the new energy reserves. Finally, those cells must have collected together into a single machine, to commit irreversibly to differentiate into various organs and at once to produce something that operates as one.

It seems that a baffling set of contingencies, unlikely ones, was instrumental in the rise of a biosphere that reached beyond the kith and kin of prokaryotes. But buried inside these events, could there have been a thread of inevitability?

The physical causes are easy to grasp. The rise in oxygen was necessary to release the power of aerobic respiration, which produces many times more energy than do other ways of making energy in oxygen-free environments. The new animals that tapped into oxygen produced more energy and more power. The rise of oxygen itself was the waste product of photosynthesis. Microbes that could grab the energy of the sun and use water as the source of electrons in this process of photosynthesis would instantly have within their reach every watery habitat on Earth with some sunlight. As a consequence, they would produce oxygen. The physical imperative and advantages to be gained by reaching into those possibilities were there. We might plausibly think that the rise of oxygen was an inevitable consequence of the evolutionary process exploring the different thermodynamically favorable energy-yielding reactions on offer on a planet.

Powering up cells by filling them with mitochondria was just physics, gathering up more power stations to produce more energy per unit volume, allowing for more energy to build more cells and more-complex structures. Any cell that evolved to be a home to more mitochondria would have had more energy available to grow and divide. This change might also appear inevitable. Endosymbiosis has happened many times during the history of life on Earth.

And what about the alliance between cells that grow to be different, an alliance that allows each cell to specialize and produce both efficiency and dedicated complexity in the tasks undertaken? This is itself easily understood when the efficiency and exquisite division of labor is translated into more successful competition by the whole organism to survive in the environment. Slime mold, an inauspicious-sounding fungus, lives a sedate life until it needs food. Then the cells gather and march in unison across the landscape, an often vivid yellow network of moving, veinlike tentacles writhing over a forest floor in search of sustenance. Over nine hundred species of these molds in far-flung corners of the eukaryotic empire show that multicellular behavior, combining to improve success, is by no means rare in the biosphere. We do not know the exact events that led cells to take up irreversibly dedicated roles for themselves, relinquishing their chance for unicellular autonomy, but even today we can see the pressure for combination and collaboration across the biosphere. On any planet where cells compete for resources and habitat, there is a good chance that multicellular behavior will eventually lead to the emergence of discrete multicellular organisms. Physical principles provide the impetus; cellular structures and their attendant genetic pathways provide the means.

Put simply, the rise of multicellularity, that is, the emergence of a complex biosphere of animals and plants, is based on uncomplicated physical principles. We can see why cells cooperated and harnessed more energy. This phenomenon would have been driven forward by competition. Large animals would have made better predators. Their prey would have evolved to be larger and thus less vulnerable. A biological arms race demanded yet more effective, and sometimes larger, machines. Once animals emerged, the vast diversity in the evolutionary experiment with multicellular creatures was assured.

There is an important distinction to be made between the claim that physical principles tightly constrain forms of life into predictable structures at all levels of its hierarchy—one that I have primarily explored so far—and the claim that the major transitions of evolution are inevitable consequences of physical principles. The second claim is one yet to be tested. Nonetheless, we can discern the reasons why life was transformed from single-celled microbes to multicellular, complex life and the physical advantages and potentialities that lay open to, and may have encouraged, life to take that route. For now, however, reducing to a list of equations all the processes that we have just been discussing would be a formidable task. To write down equations that would clearly describe the transition of a microbe to an animal is a far more ambitious task than, for example, collapsing the temperature of a ladybug into a single equation.

Considering the inevitability of these momentous transitions, given enough time for life to evolve on any other planet, it may not be wild speculation to think that life would embark on the same journey, from early molecule-encapsulating lipids to lumbering leviathans. However far these forms of life get, riven throughout them will be the physical principles that lie within the cell.

CHAPTER 6

THE EDGE OF LIFE

A VISITOR STANDS ON the pier in the breezy and delightful Victorian seaside town of Whitby, watching the seagulls on the beach pecking at bits of bread or the odd chip dropped by a summer tourist. The view couldn't be further removed from the clanking shudder of the dark cage hurtling its way down into the depths of the Earth just a few miles away.

Jump in a car, and drive north from the town, whose imposing ruined Gothic abbey was Bram Stoker's inspiration for *Dracula*, and there on the left-hand side, a perplexing sight for the unfamiliar, is Boulby mine, a collection of gray, dusty buildings surrounded by brown and white piles of salt, between them the busy roadways of a working mine. Like a cathedral to the working man and woman, nestled in the eclectic mixture of processing plants and hangars are two giant gray cylindrical towers rising into the sky, the shafts that connect the surface world with the underground labyrinth.

Visiting this hidden kingdom requires a little preparation. A bright orange jumpsuit, a self-breather in the unlikely event of fire, a hard hat, a flashlight, and, of course, a backpack full of tubes, sterile shovels, and poles—not mining equipment, but the carefully selected tools of a microbiologist seeking life deep below the surface of the Earth.

An orderly procession of miners, and their small collection of scientists behind, heads single file through the thud of large metal doors to the top of the shaft, where the cage awaits them. Messages such as "Don't be daft in the shaft" and "Act your age in the cage" scream out from the beams above the shaft. Those health and safety officials really know their stuff, and soon we are carried along by their enthusiasm ("Safety and science, it's an alliance" was the best I ever came up with).

The miners and scientists huddle into the cage, a double-decker affair, and the wire door slams shut. With a jolt, the cage begins downward. A breeze of the air ventilation system that keeps the mine full of fresh air and at relatively cool temperatures blows into our faces as the ten-minute ride to the bottom passes in darkness, the only sight the dark, salty walls hurtling past through small holes in the side of the cage.

At the bottom, a kilometer underground, we are welcomed by the familiar sight of artificial lights flooding giant cavernous rectangular holes in the salt. The salt is the remnants of a briny sea that existed 260 million years ago in the Permian period. The miners are here to drill and dig the salt, using giant automated mining machines that crunch and eat away at the rock, some of it to end up on the roads to protect you and your cars from ice and snow, some of it to end up in fertilizers to make better crops. Down here, hidden from the gaze of the public, is the raw reality of humanity up against the crust of the Earth, gathering the minerals and rocks that make our civilization work.

Spanning out, like an ant nest, are over a thousand kilometers of tunnels that Cleveland Potash Limited, the company that runs the mine, has dug since the 1970s. The tunnels, big enough to drive a van down, sprawl out under the North Sea into the rich seams of salt. Once covering an area equivalent to Europe, the ancient Zechstein Sea would have been a marvel to gaze upon. No mere brine pool, this was an enormous inland water body, its white shimmering surface disappearing to the horizon on a primeval Earth when trilobites dominated the oceans and early four-legged animals on land included the sauropsids, which would in time rise to dinosaur dominance.

As the miners peel off to the mine face, we scientists divert through some tunnels to a door in the side of the salt, like the entrance to an arch villain's lair. But behind this entrance is a laboratory that since the early part of the millennium has been at the center of the search for dark matter. Here, deep underground, where the kilometer of rock above blocks out the radiation from space, the scientists can look for the telltale signs of dark matter, that elusive component of the universe, while minimizing the interfering noise of unwanted particles streaming in from the Sun and the rest of the cosmos.

There is another reason to be here. Deep underground live microbes, denizens of the subsurface, biology's dark matter. For decades now, biologists have realized that although most of us are familiar with the trees, reptiles, birds, and all manner of other life on our surface world, a vast quantity of the mass of living things on the Earth is underground. Few animals can make a living in these depths, but many forms of energy are here, waiting to be snapped up by microbes that can flourish and multiply in the cracks and fissures of this Hades-like underworld.

In Boulby, pockets and seeps of water provide natural habitats for microbes that can eat organic carbon or munch on some rare iron compounds here and there. Small pools of water that collect in the mine provide a permanent habitat. These microbes, unlike you and me, are in no hurry to go anywhere. They have no deadlines to meet. Like life anywhere underground, they may divide slowly, and, in some underground habitats around the world, microbes may multiply once every few thousand years or perhaps longer. This is a world in the biological slow lane. From the science-fiction cleanliness of the Boulby Underground Science Facility, we head into the dark, dusty tunnels to find these seeps, collecting the water in sterile sampling tubes and returning them to our labs to extract the DNA and find out what is living in these murky depths. A diversity of salt-tolerant microbes makes a living in these austere settings: these are the *extremophiles*, literally, extreme-loving microbes.

It is sometimes said that calling these hardy little creatures extremophiles is anthropocentric. If they could observe us living in the oxygen-rich atmosphere with all those damaging oxidants so created, they would think we were the extremophiles; their nice, cozy underground home, often oxygen-free, is not extreme to them. Fun though this contrarian view is, it is actually nonsense. If these deep dwellers could express an opinion about humans walking through a rain forest surrounded by monkeys swinging in the branches, parrots squawking in the canopy, and the vast diversity of microbes inhabiting just a spoonful of rain-forest soil, they would express dismay at the decadence of the good life. There really are extremes on Earth, where physical and chemical conditions push life to its limits and where the biosphere teeters between the living and the dead. Here, animal life is excluded, and even among the microbes, habitats are occupied by just a few with the evolutionary heritage and the biochemical wherewithal to grow.

Down in Boulby, only the microbes that can take the harsh briny fluids and the sparse amounts of carbon and nutrients found here can hang on. Here, no more than a thirty-minute drive from the chatter and excitement of children licking ice creams and playing on the beach and dogs barking in the summer sun, life is at its limits. A small reduction in water availability here, an increase in brininess there, threatens instantly to extinguish it. Here, we witness the experiment in biological evolution on planet Earth as a fickle thing. Not an endless, unbounded possibility of life, powering ceaselessly through Earth, a universal phenomenon unrestricted by physics. Although tenacious, life is a phenomenon circumscribed very much in its empire, constrained like the animals in a zoo by a fence of extremes that bounds it into a pocket of existence that occupies a trifling fraction of all the physical conditions, in their extremities, to be found across the known universe.

But what are these boundaries, and are they just an unlucky and unfortunate perimeter in Earth's particular experiment in life? Would other experiments in evolution carve out new and unimaginable dominions, driving headfast into physical and chemical spaces in which our own extremophiles would wither?

With these questions, we advance to think about how cellular life is constrained not only in its construction—in the shapes of its cells and the molecules from which it is assembled—but also in the habitats it can master. Physics circumscribes the limits of life.

Boulby mine is an impressive depth, but even a kilometer is hardly a pinprick in the surface of the Earth. Go deeper, and one of the first problems for life is getting enough food. Probably only about a millionth of the space underground, all those pores and fractures in the rocks, is actually made into a home by life. The problem for microbes is not housing—it is getting energy. Life deep down is impoverished compared with the lush vegetation and relatively life-covered surface of the planet, but it is not impossible.

Dig deeper still, several kilometers deep, and now a new problem confronts life: the rising temperature. Within our planet is the primordial heat from its formation, heat trapped from the incandescent, swirling clouds of gas from which our Solar System was formed and produced from the decay of radioactive elements that form part of the recipe for making planets. At the center of the Earth, the heat is so intense that the solid iron core glows a withering 6000°C, surrounded by a liquid iron core whose churning motions, a giant dynamo, produce the magnetic field that protects us and our atmosphere from the bombardment of much of the radiation streaming in from space. Between this searing core and the relatively balmy surface, there is a gradient of heat, the geothermal gradient, and as life buries itself deeper and deeper in the Earth, it must confront this gradient.

It takes little for that temperature to rise. Even in Boulby, in the caverns and tunnels just a few meters away from the ventilated tunnels, you are hit by a stifling heat, over 30°C. Generally less than ten kilometers underground, depending on where you are on Earth and how that heat finds its way upward, temperatures exceed 100°C, the boiling point of water at sea level.

Heat up molecules in a living cell, and the bonds that hold the atoms together get so much energy that they break apart. The higher the temperature, the more damaging the energy. Increase the temperature by 10°C, and the rate of chemical reactions about doubles, so as life gets deeper underground, the rising temperature becomes ever more dangerous. It must expend its own energy to repair proteins and membranes and make new ones.

In the 1960s and 1970s, an American microbiologist, Thomas Brock, who was investigating microbes that inhabit Yellowstone National Park, wondered whether anything could possibly be growing in its boiling volcanic pools. He prodded and probed the bubbling, venting cauldrons, collecting mud to take back to his lab. Within the unremarkable sludge, he found many microbes capable of growing at temperatures of 70°C or more. This was a new and surprising discovery, and these microbes, thermophiles, did not merely tolerate these temperatures, but needed them. Cool the mud down, and they refused to grow. These new discoveries spurred others to look at higher temperatures. Like a Guinness World Records contest, pushing the upper temperature horizon of life had become something of a challenge. From the pools of scalding water at Yellowstone, scientists turned their attention to the hydrothermal vents gushing water from the Earth's crust deep in the oceans, the pressures sufficient to push the temperature of the water well above 100°C.

The hyperthermophiles, real extreme-loving microbes that prefer to grow above 80°C, cover a rich tapestry of species. Among their ranks is the record holder, *Methanopyrus kandleri*, a microbe from a black smoker hydrothermal vent, a strain of which can reproduce at 122°C.

The adaptations that all these microbes use to grow at such high temperatures bear witness to the challenge of taking on these scalding habitats. Many proteins in the cell have extra bonds or bridges of sulfur atoms, essentially bolts to hold the molecules' three-dimensional structure together against the energy at high temperatures, which has a tendency to unravel them. Proteins can be made into more-compact structures, which make them less liable to unfold. Alongside these cunning adaptations, the microbes produce heat-shock proteins, part of a network of responses designed to lock on to damaged proteins and remove or stabilize them. Chaperonins are a class of small proteins that help refold other proteins that have gone awry in the intense heat. This thermostability of the molecules of life comes at a cost. The cell must synthesize all these helper proteins, and it must make entirely new copies of proteins. This ongoing battle between damage and the energy to prevent it must set the upper temperature limit of life.

We do not know what this upper reach is. It might go above 122°C. One group of researchers suggested that a temperature of about 150°C might be the limit. With all the competing energy needs of a cell, the multifarious ways in which proteins and membranes may be made more resistant to heat, and the different sources of energy in the environment, there is not an easy theoretical way to just work out an upper limit. However, one general principle is clear. Life as we know it is based on complex carbon molecules. The various strengths of bonds between carbon and other elements are not mere serendipitous outcomes of terrestrial evolution, mere parochial numbers that exist on Earth. They are universal values. The average bond strength between a carbon atom and another, for instance, is 346 kilojoules per mole. It is this value on Earth or in a distant galaxy.

When we expose a living thing, made of chains of carbon atoms linked variously to atoms such as oxygen and nitrogen, to high temperatures, we are not confronting a chance fluke of terrestrial evolution. At the chemical level, we are challenging the integrity of the C-C bond and its universal bond energy, and other bonds besides.

At temperatures of around 450°C, most organic molecules from which life is made are destroyed. It takes special arrangements of carbon atoms to make something that is resistant to high temperature. Graphite, the material in your pencil, can tolerate much higher temperatures, but this mineral is a tedious arrangement of carbon atoms linked to one another in monotonous atomic sheets. This is not a material from which to make life forms. Place some complex organic matter into an oven at 450°C, and it will transform into carbon dioxide gas. Chemists routinely use this heating procedure to clean organic matter off their laboratory glassware. So somewhere between 122°C and 450°C, there is a hard limit to the upper extremes of life. The efforts cells expend to hold together at 122°C suggest that the limit is nearer this temperature than it is to 450°C.

I will not be so naive as to make a prediction on what that limit might be. For the point of our discussion, it is not too important. High temperature sets a border to life. Serendipitous events and chance evolutionary innovations might well modify the range of that upper limit. Maybe in the course of evolution, the development of certain heat-shock proteins shifted the upper temperature frontier upward by a few degrees. Maybe future evolutionary innovations will push life slightly higher. And perhaps synthetic biologists and genetic engineers will achieve the feat in the laboratory.

What is important is that ultimately it is hard physics that sets the upper limit, and no amount of Darwinian mutation, chance innovations, or new discoveries by life will change that boundary. Pushing the upper perimeter of temperature can certainly buy life more habitat. The typical geothermal gradient into the Earth is about 25°C for every kilometer you go down. Push the upper temperature limit by 50°C, and you have just bought yourself 2 kilometers' depth of rock where all other life has been forbidden. That is no minor amount of space from a microbe's viewpoint. It is about 1.0 billion cubic kilometers of extra rock to explore.

At a planetary scale, though, the extra temperature tolerance probably does not modify the picture of life much. Even at our conservative 122°C, the thickness of the biosphere is about 5 to 10 kilometers—compare that with the radius of the Earth, 6,371 kilometers. The biosphere is a mere 0.1 percent. Life on Earth should probably be referred to as the biofilm, not the biosphere. Even with the unorthodox suggestion that life could get near to 450°C, we would about triple the thickness of the biosphere, but its depth of penetration into the Earth's crust would still be only about 0.3 percent of the Earth's radius. Life is a thin layer on the planet, a patina of organic material denied the depths of the Earth by the limits imposed by thermal energy.

From the searing depths of the Earth, we take another trip—to the freezing extremes of the universe—and here, too, physics places boundaries on the lower-temperature capacities of life. At absolute zero, there is no possibility of molecules shifting around, linking up with other molecules, making proteins, and reading genetic codes. Trivial though the observation may be, somewhere between the temperatures you and I find clement and absolute zero, physics sets a boundary. For life, that limit is much nearer the temperatures you and I are familiar with than the chilling extremes of zero kelvin. So far, there is no good, convincing evidence of life replicating below about −20°C, although metabolic activity, the production of gases or the activity of enzymes, seems likely to occur below this.

Organisms can live below 0°C, the temperature at which pure water freezes at sea level, because there are environments where liquid water can persist. Add some salt, and the freezing point depresses to about −21°C for the lowest freezing temperature of a solution of table salt, sodium chloride. Solutions of more exotic salts, such as perchlorates, can take the freezing point well below −50°C. Part of the problem with finding active life at these low temperatures is the glacial rate of chemical reactions. Measuring growth or metabolism at low temperatures is just technically difficult.

The challenge that low-temperature life must confront is to repair molecules that are damaged when chemical reactions, and by extension many repair processes and biochemical pathways, run so slowly. Unlike life at high temperatures, damage to molecules does not primarily come from excess thermal energy, but comes from stray particles from radioactive decay; an errant zap of ionizing radiation through the cell will break DNA or destroy a protein.

All around us is radiation. Particles, including protons and heavy ions, stream in from the galaxy and the Sun. Although many of them are deflected by our planet's magnetic field, some get through and intersect with DNA.

Radiation also comes from within the Earth. In the crust or core of any planetary body, natural minerals that contain certain isotopes of uranium, potassium, or thorium produce radiation from the decay of these elements. This radiation, made up of different types of emissions, including damaging gamma rays, affects us all, although it is of sufficiently low intensity that we rarely worry about it.

This *background radiation* causes damage to the major components of life, particularly DNA. Background radiation can break the double helix of DNA or cause the formation of reactive oxygen radicals, which themselves attack and damage the DNA molecule.

It can disrupt any sort of complex, long-chained compound from which life might be formed. There is no easy way for life to screen this radiation. Most of it can effectively penetrate biological material, a problem that is exacerbated if you happen to be a microbe made only of a single cell. Your only option is to repair the damage once it has happened.

Added to these problems is the natural tendency of some molecules to alter their structure, or decay. Sit there and do nothing, and all this damage will slowly, steadily accumulate until, eventually, it is so great that the cell will never recover. At some low-temperature region of active life, there must be just enough energy and biochemical activity for the cell to repair this unavoidable destruction, but not so little activity in the cell that it cannot keep up with damage collected in molecules over long periods. Like life at high temperatures, there is no easy way to calculate unequivocally that temperature. We know, however, that cells can evolve stunning ways to deal with plummeting temperatures. The so-called psychrophiles, microbes that grow best at below 15°C, have developed ingenious methods to offset the effects of cold.

At the freezing temperatures found at the Earth's polar regions, life's membranes, made of lipids, freeze solid. Place a chunk of butter in a warm kitchen, and it is soft, easy to spread on your toast. After eating your breakfast and putting the butter in the fridge, you will notice that once the butter has sat there for an hour or so, it goes solid. Like the fatty acids in the butter, the fatty acids in a microbe's lipid membranes, in the refrigerated temperatures of the Antarctic ice sheet, freeze solid. The lipids contain long chains that, when cooled, line up side by side, serried ranks of molecules that pack together tightly, providing little room for movement.

There is a way to make them more mobile. The fatty acid chains comprise many carbon atoms attached to one another by single bonds, one carbon after another. Introduce a double bond into the chain, and you create a kink. The molecule, rather than being a long straight chain, now kicks out to one side. Lay these unsaturated fatty acids, as they are known, side by side, and they will not pack together. Because the rogue chains keep the fatty acids separated, they can now move around more freely and are less able to pack up tight in the freeze.

Back in the kitchen, we can illustrate this idea with another common item. Safflower oil is full of unsaturated fatty acids. Put it on the kitchen table, and it is runny. Place it in the fridge, and it remains liquid. Unlike the fatty acids in butter, the oil's kinked chains keep the molecules moving. In just the same way, microbes in the polar regions fill their membranes with unsaturated fatty acids. These branched molecules keep the membrane flexible and fluid. When exposed to subfreezing temperatures, the membranes, instead of freezing solid, now retain the malleability that allows food to move in and waste to get out of the cell.

This example is one of stunning beauty. By the mere switch of a single bond to a double bond in a chain of carbon compounds, life can now colonize the frozen wastelands of Earth. Within this modification at the atomic level, we see how evolution can use simple tricks of chemistry and physics to master entirely new worlds.

Yet magnificent though this simple invention may be, its accomplishment is to help push the lower limit of life down toward the hard barrier set by physics. At that lower perimeter, no change in membranes, no mere alteration in a few bonds, can escape the twilight zone, where even the most energy-rich environments are trapped in a chemical slow lane. In this zone, reactions are so tardy that no living entity can do chemistry fast enough to keep up with the inevitable assault of damage, molecular transformation, and disassembly.

Thus, life is trapped between a hot and a cold place, its boundaries of operation constrained by simple principles. Invention and chance may well vary the exact limit at which life operates in any time or place in which it finds itself, but these limits are narrow. Even with wildly optimistic assertions about its temperature tolerance, the percentage of a planet that life can master is a tiny proportion of the total volume. Changing its capacities by hundreds of degrees merely alters the volume of a planet it may occupy by a few tenths of a percent. Life is hardy and persistent once it gets hold of a planet, but its dominion is small. Between absolute zero and the temperature of the interior of a star, say, the Sun, life occupies only

0.007 percent of this temperature range. This narrow vista is not a consequence of historical facts of evolution it might break free from if the evolutionary process were run again. It results from the laws of physics operating on a small but interesting branch of chemical compounds we call life.

Across the universe, and even on a small rock like ours, life meets a bewildering variety of other extremes. Perhaps temperature is an anomalous example, an unusual set of physical boundaries. Do other extremes narrow the range of life?

Travel to Guerrero Negro, in Baja California Sur, on the west coast of Mexico, and there you will find the biggest salt works in the world. This large, flat area near the coast contains thirty-three thousand hectares of salt fields, enormous expanses in which seawater is evaporated under the careful watch of its owners to make pancakes of white salt, glittering, dazzling in the sun. Each year, this industrial-scale operation makes nine million metric tons of salt.

Like the salty ponds deep in Boulby mine, life here has a huge challenge. It must deal with the process of osmosis, which relentlessly causes the salty environment to suck water from the cells, transforming them into puny microscopic prunes. No random evolutionary adaptations can simply escape osmosis. It is an inescapable process, and wherever it exerts its effects, life must adapt or die.

No one should be surprised to learn, then, that like thermophiles, these regions of the world contain life that has adapted to such an extent that it now needs high salt to survive. The halophiles, microbes that need between 15 and 37 percent salt to grow, inhabit Baja and Boulby and thrive in waters that would make most life shrivel up and collapse.

Osmosis will not just go away. Its effects are uncompromising. It can be stopped only by exerting a pressure, the osmotic pressure (π), against it equal to:

$$\pi = imRT$$

where m is the number of moles of the substance per liter (the molarity), R is the universal gas constant, and T is the temperature. The value i is the curious van 't Hoff factor, which has to be worked out using experiments. It is the degree of dissociation of the salt ions added to the water.

The osmotic pressure that you must exert merely to stop pure water from being sucked up by salty seawater across a membrane nearly equals a crushing twenty-eight atmospheres of pressure.

Faced with the trauma of having water extracted by force, life has evolved two effective ways to respond. You could just take up ions into the cell. The so-called salt-in microbes took the approach: if you can't beat osmosis, join it. By allowing potassium ions to accumulate in the cell, the osmotic potential on the inside and outside of the cell is equalized and the cell can continue life as normal. The side effect is that the cell now has a high concentration of salty ions inside, and it must evolve its proteins to deal with this problem. Ions can disrupt bonds, interfere with the folding of proteins, and change the availability of water.

Evolution's response to this problem is ingenious. Many proteins have within them hydrophobic portions. As salts will displace water, these water-hating regions of proteins become even more attracted to each other as the water between them is pushed aside by the salts. This increased attraction is a problem because now these protein units may be too stuck together. By evolving proteins whose hydrophobic contact areas are reduced, the connection is made a little weaker, compensating for the salt that strengthens it. This subtle little trade-off brings the proteins back into normal operation. By changing charges and bonds in the key proteins of life, other adaptations add to the repertoire of modifications to live in salt.

However, some life forms eschew salt completely. Instead, to maintain the osmotic balance, they produce compounds that act like salts but that are slightly kinder to the cell. Sugars such as trehalose and some amino acids can increase the concentration of compounds inside the cell so that the osmotic pressure equalizes with the outside without the damaging effect of salt ions. These "salt-out" microbes are very common and inhabit the salt crusts of Baja and the brine seeps of Boulby.

Push the salt concentrations high enough, and eventually the problem is not so much all that intense osmotic pressure, but now just a general lack of water altogether. The cells become so depleted in water that this essential solvent for life, the liquid that makes the chemistry of life possible, is no longer in enough supply to maintain the machinery of life in a working state.

This lower floor of water stress is set by the water activity (a_w), a measure of the availability of water. In more exact terms, it is the ratio between the water vapor pressure above the salt and that of pure water. The smaller the water activity, the less water available. Pure water, distilled water, has a water activity of 1. A saturated salt solution, the sort of briny pond in Boulby, has a water activity of 0.75. Most microbes need a water activity of about 0.95 or higher; any lower than this, and osmotic stress sets in, and the molecules of life cease to function. However, the halophiles and many other microbes that can tolerate dry environments can push that limit much below 0.75. Some fungi can reach a water activity of just below 0.6.

Like scientists enamored with finding the temperature extremes of life, there is a quest to find the water activity limit of living things, and no doubt with enough scurrying around the world and digging around in extreme environments, we'll see that limit drop. But from the viewpoint of this book, the rat race of microbiologists interested in defining the new limit is less important than the more general point that water availability restricts life. Once the availability of that most fundamental of life's requirements, liquid water, is pushed below a water activity of about 0.6, the diversity of living things that can persist in that realm is diminished. By a water activity of about 0.5, there are unlikely to be any active living things. There just is not enough water. A familiar culinary analogy is honey, whose water activity is usually below 0.6. That is why you can leave this sweetener on the kitchen table without its going moldy. Honey is a parched desert for a microbe.

Water activity, and, indeed, honey, shows us that some places that contain liquid water are uninhabitable. We are all generally used to thinking of any watery environment as hospitable to life. You will often hear planetary scientists say that the search for life on other planets is about "following the water." This is sometimes colloquially said as "where there's water, there's life." The aphorism comes from our everyday observation of the essential role that water plays in living things, but you can see that this claim is not 100 percent accurate.

Aside from honey, there are other watery solutions too extreme for life. A saturated solution of magnesium chloride at 25°C has a water activity of 0.328, well below the preferences of biology. These solutions can also cause disorder in biological molecules. Therefore, even on Earth, we can find places, such as deep brines in the Mediterranean, that contain biologically dangerous levels of magnesium chloride. When investigated by microbiologists, these brines are found to be at the limits of life.

Deep in Boulby, channels of water cut hither and thither, dissolving their way through sodium chloride here, making their way through sulfate salts there. Throughout the mine, almost all the brines contain active life, halophiles making a living on their depauperate resources. Sometimes, these rivulets of salty water cut through a vein of magnesium chloride, and when they do that, the water activity plummets. Within these small trickles of water, there is no evidence of life. A small detour through a vast sequence of rocks, through hundreds of meters of salt that is extreme but benign to salt lovers, into one particular salt pushes life beyond its capacities.

Similarly, Don Juan Pond, an undistinguished-looking water-filled hole in the McMurdo Dry Valleys of Antarctica, is thought to be devoid of active life. Since the 1970s, scientists have been intrigued by this rare water hole. Because it is filled with a brine of calcium chloride with a water activity of below 0.5, we would expect the hole to be lifeless. Indeed, microbiologists have had mixed results in growing things collected from Don Juan. The consensus is that the pond is empty of active life but that microbes washed in from the outside remain viable. When fished out and plated onto agar plates in more benign laboratory conditions, they will grow. It is quite possible that in the spring, when snow melts in Antarctica, the water flows into the pond, forming a lens of freshwater on its surface, where the water activity is above the threshold for life. Then, life may briefly take a respite and multiply before the pond is mixed and it returns to its uninhabitable state.

These lifeless but watery habitats show us something profound and important. Water is a requirement for life. However, even on this planet, we have environments where there is plenty of liquid water but its availability to life is insufficient, not merely because the environment is dry, a solidified salt crust in the Baja Californian sun, but because even in a liquid state, some salt solutions fail to relinquish the water molecules needed for life. We need not visit alien worlds to find where life has reached its physical limits, where no amount of chance or evolution will push it beyond the barrier of salt. For over three and a half billion years, evolution has been experimenting with adaptations, yet when faced with a low water activity, evolution is impotent. You might fill a saturated magnesium chloride or calcium chloride brine with nutrients, organic material, and every conceivable energy source you can imagine, but still that watery environment will remain dead to multiplying cells.

Startling though these restrictions are, let us continue to other extremes to see what else might stop life in its tracks. We need to explore some other extremes to construct a general picture of how physics might bound the biosphere. Travel to the south of Spain, near the ancient and architecturally stunning town of Seville, and you will come across the Rio Tinto, a bright orange and red river that cuts through the Iberian Peninsula. Flowing for more than a hundred kilometers, the river slices through a belt of sulfide rocks that oxidize to make sulfuric acid. The result is a highly acidic river, with an average pH of 2.3. Even this level of acidity is tame compared with Iron Mountain in California, where similarly acidic streams have a pH of 0 to 1, as low as battery acid. Considering such extreme chemistry, we might be forgiven for believing that conditions would again be too extreme for life.

Yet we find life thriving in these locations. The pH of water is a measure of the concentration of protons. The more protons, the more acid the solution. Protons are not a bad thing for life. The flow of them through the machinery of the cell membrane is the basis of energy harvesting. However, if there are too many, the charge they build up will damage proteins and other crucial parts of the cell. The acid-loving microbes that live in the Rio Tinto and Iron Mountain, the *acidophiles*, have to work hard to keep those protons out, and they do this by pumping them from the cell to keep the interior of the cell at a near constant, almost neutral pH. To call them acidophiles is something of a misnomer, as the microorganisms have evolved to keep their cell interiors from becoming acidic, but despite their exhaustive efforts to keep protons out, they are adapted to these conditions. Place them in a less acidic environment, and many will die.

At the other extreme are the alkaliphiles, microbes that can tolerate high-pH environments. A trip to Mono Lake just north of Death Valley, California, will give you a glimpse into the world of alkaline life. Here strange tubular carbonate mounds, called *tufas*, climb from the lake and the land around it in an eerie alien scene, chimneys that are testament to mineral precipitation in a lake with a pH of 10 and a saltiness three times that of the ocean. This high pH is no barrier to life. Not only do microbes grow in the lake's waters, but alkali flies (*Ephydra hians*) run higgledy-piggledy along its shoreline, while brine shrimp (*Artemia monica*) flex and pulsate in its waters. Here, even animal life can thrive. The fly larvae, which begin their life in Mono Lake, have within them special organs that turn the alkaline waters into carbonate minerals. You can think of these biominerals sequestered in the larvae as a detoxification method, a clever way of removing the ions in the water and collecting them into minute grains and out of harm's way.

Mono Lake, although a source of fascination and a focus for many scientists, is not the most alkaline lake in the world. Other lakes around the world, such as Lake Magadi in the Rift Valley of Africa, with a pH over 11, also host ecosystems.

So far, we know of no natural extreme pH environment that precludes life on Earth. Have we now found an extreme in the face of which life defies physics? Well, not really. We might appreciate this by thinking about the physical facts. High temperatures must eventually exclude life since in extremis, life is destroyed by the vast energies injected into the atomic bonds of its molecules. A fragile carbon-based life form just cannot hold those molecules together at 1,000°C, so we can easily and simplistically grasp the likelihood of an upper temperature extreme for life, although we can investigate where that limit might be and what molecular failure ultimately defines it. Similarly with salt. In the simplest version of our understanding, the limit of salt or desiccation tolerance is set by the availability of water. Remove the water entirely or add salts so that the water molecules are all but unavailable, and life is denied the solvent it needs to operate. An edge of existence based on the availability of water is also easy to grapple with.

When we turn to pH, there is nothing inherent that will shut down life. As long as cells have enough energy and good enough pumps to either remove protons from the cell or let them in, then the interior of the cell will remain near a neutral pH, undamaged by extremes of pH in the outside world. The ions themselves, provided they remain outside the cell, present no mortal threat. Perhaps not surprisingly, the different pH environments explored so far contain life.

Not that pH is always kind to life. Add it to other extremes, such as high temperatures or salt stress, and the cell must now find enough energy to deal with multiple problems. In most of Earth's environments, there is rarely just one extreme, although there are many environments where one will dominate. In the deep oceans, cold temperatures combine with saltiness. In volcanic pools, acidity often combines with high temperatures. Microbes have been found that can cope with salt, high pH, and high temperature. In any environment, an extreme might push life over the edge when the cell lacks sufficient energy to cope with the onslaught from a mishmash of them. Yet, on its own, pH does not seem to be a fundamental limit to life in the Earth's natural environments.

Other extremes too have no known ultimate limits. In the crust of the Earth and deep in the oceans, high pressures can contract and constrict the molecules of the cell, yet life is found at the bottom of the Mariana Trench, some eleven kilometers deep in the oceans, where pressures are a thousand times higher than at sea level. In the crust of the Earth, deep underground, life thrives. Adaptations are found to deal with these elevated pressures. Pores and transporters across the cell membrane help expel waste and take up nutrients; proteins are modified in these so-called barophiles, or pressure lovers. As you dig down into the crust, temperature will most likely limit life before pressure does, the geothermal gradient exposing life to prohibitive heat before the effects of pressure can immobilize a cell.

Here again we have much to learn. In the absence of the temperature problem, does life succumb to some pressure limit? The problem with high pressure is that it indirectly affects many other things, such as the solubility of gases and the behavior of fluids. Whether life at pressure extremes would ultimately give in because of the direct effects on the cell or because the changing behavior of its milieu might starve it of nutrients or energy are matters still to be fully studied.

Among the plethora of other extremes, one extreme might pose a hard limit to life. Ionizing radiation, like high temperatures, imparts energy to biological molecules and damages or destroys them. We know that life can resist the effects of radiation. Molecules such as DNA can be repaired where the strands are damaged. Proteins can be constructed again, and some pigments, such as the carotenoids, can quench the reactive oxygen states produced by the radiation as it slams into water. Life has an armory of responses for dealing with molecular damage caused by radiation, and when these responses are assembled into a single microbe, the results can be impressive.

The humble *Chroococcidiopsis*, a cyanobacterium that lives in rocks in the world's deserts, can take a dose of about fifteen kilograys, about a thousand times the dose that would kill a human. This microbe joins the ranks of *Deinococcus radiodurans*, another bacterial life form that, through repair and damage mitigation, can hold off ten kilograys of radiation or more.

There must be an upper limit to radiation. Assault a cell with enough of this form of energy, and the ability to repair and make new molecules will be overwhelmed in a similar way to the destruction caused by high temperatures. On our planet, where only a few natural and artificial environments expose life to sustained high doses of radiation, excesses of radiation have likely not confronted evolution with a hard limit as often as, say, extremes of temperature have done. Nevertheless, we might imagine that such a boundary exists.

The biosphere is like a zoo, surrounded by a wall. Within it, all manner of living things, minuscule and giant, have evolved, guided by laws into predictable forms. Restrictive though these rules are, they permit the burgeoning of an experiment in biological complexity that is extraordinarily diversified in its minutiae. However, the biosphere's potential is ruthlessly curtailed by the tough perimeter surrounding the zoo. Some of these limits are probably universal. No evolutionary roll of the dice can overcome a lack of a solvent within which to do biochemistry or the energetic extremes of high temperatures. The details, the temperature sensitivity of this and that protein, may well modify the exact transition between the living and the dead, particularly for individual life forms, maybe for life as a whole. But in broad scope, life's boundaries, the insuperable laws of physics, establish a solid wall that bounds us all together.

This zoo is by no means expansive. From a cursory survey of the kaleidoscope of life on Earth, it is easy to think of life's diversity as endless, and its small variations are. The physical space that life occupies at the planetary scale, and the physical and chemical conditions it can adapt to, within the vast range of conditions found across the known universe, are petite. We live in a diminutive bubble, circumscribed by universal extremes, within which the restricted trajectories of evolution explore their reach.

THE CODE OF LIFE

"WE HAVE FOUND THE secret of life!"

These immortal words were uttered in The Eagle, a pub on Free School Lane in Cambridge, England, the day the structure of the genetic code, DNA, was discovered.

I suspect what was actually said was, "Jim, what are you having?" "Oh, a pint of lager please, Francis," "OK. That'll be a pint of lager and a Guinness, please, and two packets of pork scratchings." Or something like that.

Anyway, far be it for me to ruin the romantic dreams of the credulous. But I have little hesitation in saying that in February 1953, when James Watson and Francis Crick, with inspiration from X-ray images made by Rosalind Franklin, proposed a structure of DNA, a monumental step forward was made in deducing the centerpiece of life. This molecule is the code that contains the instruction manual from which living things on Earth are made—the cipher of the cell.

When the secrets of this molecule were unraveled, it was not surprising that those who surveyed its features would, for many years to come, consider DNA a freak, a chance product of evolution—a molecule whose structure was special. Its architecture seemed so unlikely that if anyone was asked whether such a molecule could evolve on another planet, they would probably have answered that it wasn't impossible, but that an event of such low probability would be astounding. In an early paper on the evolution of the genetic code, Crick himself described it as a "frozen accident" that occurred at the birth of life, got fixed in the very bedrock of living things, and, once there, could not be displaced without a catastrophic effect on the cell, probably causing its death. Once such a crucial code and its entourage of structures to read it were in place, the smallest error or alteration would be fatal. This view is compelling, but seems increasingly unlikely.

At this next level down in the hierarchy of life, from cells to the molecules that encode and fabricate their form, new light has been shed on evolution's choices. Here too we have begun to see the indelible mark of physical principles operating through the chemistry of living things, channeling and cajoling the code of life into an edifice that seems to have very much more to it than the mere quirks of chance.

Take this molecule, a double helix. Unwind it flat, and place it down on the table in front of you, magnified. On your left and right are the two backbones running down the table toward you. Made of repeating units of phosphate and ribose (a simple sugar), these two chemical struts of the DNA ladder hold the whole molecule together. Between them, running down through the middle, are the guts of the machine, the rungs of the ladder. Sequentially attached to the backbone on the left and right is the four-letter alphabet of the genetic code, made up of adenine (A), thymine (T), cytosine (C), and guanine (G). Strung along, one by one in endlessly varied combinations, these nucleobases spell out the information that the cell will read to grow, repair, and build copies of itself.

These four little molecules are peculiar, for they bind to other members of the group in a very specific way. An A can only bind with a T, and the C only with a G, and vice versa, to form a pair. As the letters of the code have this very fussy binding preference, if there is an A on the left, then there will be a complementary T on the right, and so on. Traversing through the center, attached to the two backbones of DNA on either side, are these *base pairs*, A's and T's linked together, spiraling down the center of the double helix, intermixed with pairs of C's and G's.

Right here, we have the first suggestion of something strange. Few molecules in nature have this tendency for a very particular ability to lock onto another to form such a small and tight little family of structures. A freak accident, it seems.

This apparently odd property was not lost on Watson and Crick, who observed in their paper describing the assembly of DNA, "It has not escaped our notice that the specific pairing we have postulated immediately suggests a possible copying mechanism for the genetic material." Pull the two chains of the DNA apart down the middle, and now it is a relatively simple matter for the cell to produce two copies of DNA, since the one single strand can be used to resynthesize the other one. It knows that any A must be bound to a T, and C to a G. So two single strands can be used to make two new double-stranded DNA molecules.

At the core of the code are those four chemicals, the four letters of the alphabet, A, T, G, and C. Surely it is just chance that the code has a magic number of four? Why not two, six, or eight?

Long before the emergence of life, scientists think that the world was monopolized by RNA, the close relative of DNA. To this day, RNA acts as the intermediate between the code in DNA and functioning proteins. Slightly more reactive and less stable than DNA, the RNA molecule has the remarkable ability to fold in on itself and, like proteins, to make active molecules that can catalyze chemical reactions and even replicate themselves. In this "RNA world" over four billion years ago, self-replicating molecules were dominated by RNA, with proteins attached here and there. Eventually, the sequence of letters in RNA, by some trick we do not yet know, is thought to have been coded into the more stable DNA molecule that today stores genetic information during the cell cycle.

Imagine a genetic code with a two-letter alphabet, say just C and G, the entire genetic code just a long Morse-code-like iteration of these two chemicals. In the RNA world, these two bases could bind as they do today to form C-G base pairs, allowing for the RNA molecule to fold and make complex shapes that can multiply and carry out chemical reactions. However, the binding is not very particular. Each base has a chance to bind with 50 percent of the other bases in the code (assuming they are split fifty-fifty in abundance, C's might bond to any G's and vice versa), making the association between bases rather unfussy. Now add another two bases—A and U (uracil, which replaces thymine in RNA)—to make four, and we have something that can do more-complex binding tricks and contains more information, more complexity. Each base can now link only with 25 percent of bases, and the structures can be more refined. Pairing becomes fussier and the molecules more intricate. In essence, the more bases there are, the more information you can have in a molecule or the shorter the molecules can be with the same information.

However, push that number beyond four, say to six or eight, and you have more information, but now there are other problems. As you add more bases, it gets more difficult to find ones that are sufficiently dissimilar to make it easy for them to be distinguished when the molecule replicates. One consequence is that the rate of errors is usually higher and mismatches are more common when the code multiplies. Computer modeling of these early replicating entities suggests that like Goldilocks making a genetic code, the number of bases in the real code, four, is just right.

Other lines of evidence have converged on the same conclusions. Studies using computer models of RNA molecules virtually multiplying and changing have found that of the various numbers of nucleobases possible, the use of four gives the molecule the greatest fitness and the greatest ability to evolve.

As with many of these ideas, they somewhat flounder because we do not have a time machine. Did these molecules replicate as we imagine they did on early Earth? Was there an RNA world, and was it really as we imagine? No one would claim that we have a conclusive answer, but there is something uncanny about the outcome of these different tests, that when we do them, we do not arrive at the startling realization that life on Earth probably has it all wrong. We do not discover that life would be much more effective and more likely to have evolved more efficiently if it had a code with a different number of bases. Instead, we keep coming back to what we observe in the structure of our own biology.

This conclusion does not rule out the possibility of frozen accidents, chance evolutionary paths locked into the structure of early
life that, once there, could not easily be altered. Furthermore, the
findings rely on a dim and distant past. Many of the hypotheses for
why the code has four letters are rooted in a supposition about an
RNA world, a four-letter advantage with its heyday in a world now
long since gone. Despite these limits in our knowledge, the research
suggests nonrandomness. The code of life and the way it is read are
probably not mere contingency, one of a plenitude of possible paths.
Instead, many of its routes and diversions, experiments and errors,
ultimately led it to a structure that is predictable and congruent with
physical processes and rules we are beginning to fathom.

The number four may have some significance, but surely the
chemicals themselves could be anything? Surely what matters is that
they are different and therefore by simply stringing them along a
chain in different combinations that can be read, we can produce a
diverse code with many permutations of "letters" to build the things
needed to construct a life form?

Since the turn of the century, extraordinary advances have been
made in modifying the natural genetic code. Motivated by a desire to
"expand the alphabet of life," as it is sometimes referred to, synthetic
biologists can produce genetic codes with more than four letters. A
larger alphabet would allow them to pack more information into the
code (accepting that this increase can lead to more errors in replication) and to experiment with making cells that would produce
new drugs and other useful products. This motivation has forced
synthetic biologists to try to discover how the structure of the code
evolved and whether different chemistries are possible. Could the
code be something else among many possibilities?

Laboratory investigations with alternative bases that share
similar chemical structures with the known code but with slightly
different configurations of atoms have thrown up a series of other potential choices. The unwieldly named xanthosine and 2,4-diamino-
pyrimidine couple is one such base pair. Isoguanine and isocytosine,
which share the same chemical formula as the traditional bases of

guanine (G) and cytosine (C), but which have some of their atoms flipped into different positions, is yet another. Some isoguanine and isocytosine can even be incorporated into cells that can be deceived into replicating with these alternatives added into their DNA.

Experiments like these show that nature can use different codes, but to explain why nature picked the bases it did, researchers need to methodically try out all sorts of chemicals. Scientists at various institutions, from the Scripps Institute for Chemical Biology and Harvard University in the United States to the Eidgenössische Technische Hochschule in Switzerland, have painstakingly looked at pairing in many possible bases in RNA. Their work is like a journey across a landscape of chemicals, prodding around to see whether going in different directions makes any difference to pairing.

They tried to make RNA out of bases made of hexopyranoses, which share a chemical similarity to our familiar bases, but hexopyranoses are made of a six-carbon ring and not a five-carbon ring and are slightly larger. This greater size hinders them from forming a proper pair. Only in one instance when some of the chemical groups (specifically, -OH groups) were removed from one of the rings could base pairing occur, but this is not a likely natural chemical to be found in a genetic code. With this result alone, the research shows that the four letters chosen by life are not random, but the makeup and arrangement of atoms play an important part in how a genetic code can be assembled. Make the molecules too big, and they will not pair up.

On the chemists slogged, into wider territory. When they made isomers of RNA, where the chemical structure was the same but the chemical groups were attached to different positions, such as the laboriously named pentopyranosyl-(2'→4') systems, the researchers made new base pairs that worked. Somewhat remarkably, some pairings are even stronger than those found in natural RNA. Does this suggest that here, in a backwater of the nucleic acid world, was an undiscovered set of compounds that would make better bases than the ones taken up by life, a small oasis of chemistry more propitiously placed to provide the key components of a genetic code?

One feature of nucleic acids is flexibility and the possibility of opening and closing base pairs to replicate or read the code into protein. In the hypothetical RNA world, base pairs had to be sufficiently strong so that a molecule would remain folded into the right structure, but sufficiently weak that it can be flexible, allowing for that folding in the first place. In this outpost of structures that have stronger binding than natural RNA, we probably have molecules that are too inflexible. Consequently, RNA might not have been made better had it used these other bases; its structure and choice of bases reflect an optimization of base pairing and not a maximization of the strength of base pairing.

Synthetic biologists will doubtless have much more to say about the choice of life's genetic alphabet as they continue to try new chemical arrangements. This work will ultimately shed light on the fundamental and ancient choices made by evolution in building an information storage system. But for now, we can say that the choice of chemicals in this code seems to be determined by simple physical mechanisms.

Onward with our exploration, we might wonder whether reading the code into something useful entails more chance, less room for predictable physical channels. The first stage of reading the genetic code is to render a complementary copy of the DNA in RNA. Not surprisingly, this strand is called *messenger RNA*, for its task is to be the proverbial messenger, a complementary copy of the long DNA code that can be carried away and turned into the final product, protein. This RNA copy of the DNA code is synthesized by RNA polymerase, a large enzyme that ratchets along the DNA, binding bases together to slavishly create our tentacle-like messenger.

Along the length of this messenger RNA, bits of yet another RNA molecule assemble. These bits, called *transfer RNAs*, bring to the strand their little cargos of amino acids, the component blocks of proteins. Each transfer RNA has its own special amino acid, and each prefers to bind to a very particular part of the code.

The transfer RNA must bind to three letters of the code, a so-called codon. As the transfer RNAs shuffle and snuggle up to each group of three consecutive letters along the messenger RNA strand, their passenger amino acids come into contact and bind to one another to make a chain of amino acids. From this machinery, itself protected by the ribosome, a collection of giant RNA structures, a strand of amino acids appears, like a snake emerging from a hole. Once released from the ribosome, the long chain of amino acids will spontaneously fold together in intricate contortions. A protein has been formed. This newborn molecule is ready to carry out a chemical reaction, participate in building a membrane, or do one of the numerous things required to build a self-replicating life form.

From DNA to RNA to protein, reading the code is, on one level, a process of refined simplicity. First, DNA's four letters read into a message, then the transfer RNAs bind to this message with their amino acids, themselves having a striking preference for a three-letter code. Finally, a string of amino acids is fabricated, the protein itself. On another level, the process seems a thing of tortuous particularity. Four oddly self-binding chemicals come together to make a code from many millions of natural chemicals in the environment. To cap it all, the proteins that emerge from this code are made of just twenty amino acids, while many hundreds of these molecules exist in the natural environment.

Let us recap what we need in our information-of-life toolbox. Between RNA and DNA, we have five major bases (A, T, C, and G in DNA, and A, U, C, and G [the T swapped with U] in RNA), a backbone made of just phosphate groups and ribose sugars, some transfer RNAs (actually, the cell requires at least thirty-one such RNA molecules to read the code), and twenty amino acids. (Some cells use two other amino acids, selenocysteine and pyrrolysine, bringing the total in life to twenty-two.) We have an entire information storage system that can go from a code to production of functional molecules with fewer than sixty molecules. There are two ways you can look at this. Either this system is a chance event, an accident of absurd proportions that could have happened many other ways, or evolution has been very selective; very few paths, perhaps even just one, were possible. Perhaps these sixty or so molecules are special in the pantheon of organic molecules that exist in the known universe?

Answering this question is perhaps one of the most profound challenges that has confronted biologists since the code was unraveled. Its answers would establish whether the structure of life's code and its products are pure chance or whether deeper physical principles have shaped them.

Having built a four-letter code, we now have the mystery of how that code was assigned to the different amino acids. As described, three consecutive bases in DNA code for each amino acid. Now as each of these bases could be one of the four letters, A, C, G, or T, that means we have $4 \times 4 \times 4 = 64$ possible combinations. However, life needs codons for only 20 amino acids and sometimes 22. So what gives? Many of the amino acids used in life have more than one codon assigned to them. This *degeneracy*, as it is called, leads to a table of all 64 three-letter codes assigned to different amino acids. Within its ranks are two punctuation marks—a *start* and a *stop* codon—that tell the code when to begin and end reading, the markers that define the start and end of a gene. Each gene codes for a part of a protein or a full protein.

This table of codes, showing the association of each three-letter code to its respective amino acid, is a little like a Rosetta stone. Besides modifications here and there that lead to about twenty modifications to this core table, the layout is universal across life. This universality speaks of an ancient origin for the assignment of the three-letter codes to their respective amino acids. It suggests that an ancestor to all life on Earth today contained this code, which was then propagated during evolution to all living things. Fathoming out how this table came about and whether it too might be a freak accident has occupied the minds of inquisitive scientists. Regardless of who is right, most scientists, at their core, propose that the table is not a random accident, but that it is the result of quite specific selection.

One crucial thing a life form wants to do is prevent too many errors in the reading of its genetic code, either when the code is being copied or turned into useful proteins. Perhaps the organization of the amino acids with particular sequences of the code is a way to minimize the chances of these errors creeping into proteins.

Curiously, codons for the same amino acid tend to be bunched to-gether. The codons for the amino acid alanine, for instance, are GCU, GCC, GCA, and GCG, where only the third place varies. This same pattern applies to other amino acids such as glycine and proline. By bunching codons together like this, the chances are that a small error in the code will not alter the amino acid, leaving the final protein unchanged. These accidental revisions in the code might come from a mutation in the code itself, maybe caused by radiation or a chem-ical modification to the DNA, or a mistranslation of the code when the messenger RNA was being read. Either way, the code, so mani-fested, reduces the impact of errors from wherever they may come. Furthermore, amino acids with similar chemical properties seem to share similar codons, which has been explained as a way to minimize the impact of a mutation or misreading in the DNA on the protein eventually made.

If you run a computer program to compare genetic codes that are more efficient at reducing the chances of mistranslation, then the natural code appears to be very unusual. Of all the codes that nature might make, out of a million alternatives, our own code was the best at reducing these errors.

There is another tantalizing clue behind nature's choices in that table of codes. The amino acid arginine happens to be able to bind to the codons to which its transfer RNAs can also bind. The same is found for the amino acid isoleucine. This suggests to some people that the codon table results from ancient attractions between amino acids and little strands of RNA, perhaps even before transfer RNAs became the intermediary between them. Perhaps amino acids bound directly to the messenger code without all that complex machinery we see in operation today. Those affinities laid the groundwork for the link between the decoding of RNA into proteins.

It is easy to get sucked into polarized arguments, but when all the possibilities are considered, we might expect that maybe el-ements of all these theories are built into life. When the first code emerged, it seems logical to suspect that certain amino acids bound to bits of RNA and these affinities may well have something to tell us about how certain codons wound up coding for particular amino

acids. And perhaps, parallel to these developments, evolutionary selection would have favored codes in which errors were minimized
or at least reduced to levels sufficiently non-deleterious to allow for
reliable reproduction. The fewer the errors, the more likely that offspring molecules would function properly and be propagated in the
environment. Later, mutations may have led to reassignments of the
codons to optimize the code further.

There does seem to be a little conundrum in all this. If the table
is so crucial to life, such a core part of the genetic apparatus and its
translation, surely once it got stuck into the very earliest life, it would
remain there as, in Crick's words, a frozen accident? Surely then, we
should expect it to be a highly imperfect thing, full of idiosyncrasies
that are a shadow of its early and manifestly vital part in the information storage system of life, a system whose later alteration would
spell death for an organism? However, the adeptness of synthetic
biologists to reassign codons to completely new amino acids in the
laboratory shows that life may have more opportunity to experiment
than once was thought. There exists room for change. In the natural
environment, there are ways these swaps in the table of codons could
have happened even after the fundamental architecture of the genetic
code was established. A cell might stop using a certain codon. Perhaps a mutation made it lose the gene for a transfer RNA and thus its
associated amino acid. Then later, through a duplication of another
transfer RNA gene and its mutation, the codon could be reassigned
to a new amino acid entirely. Through such genetic reassignments,
the table may be changeable. Like metabolic pathways, it seems that
life can traverse new paths and try new experiments in codes.

This flexibility in the biochemistry of life has a much more fundamental general implication—that historical chance, the contingent
roles of a pair of dice, may not get locked into life as frozen accidents and immutable legacies of history as rigidly as has always been
assumed. If life is somewhat malleable in the way it can change its
molecular machinery, then life can also be shaped by physical principles, to be optimized against the laws of physics, and it is not ines-
capably imprisoned in a molecular straitjacket wrapped around it at
the dawn of life.

The enduring question is to what extent this flexibility of life's biochemistry leads to predictability. Would some alien, with no prior knowledge of Earth, but with some rudimentary knowledge of biological information storage, be able to predict a priori what we now observe—four particular bases and a codon table that translates the code?

We probably still have much to learn about the flexibility and evolution of the code to answer this question. Synthetic biologists will take us closer to a better understanding. However, that the code is just an accident, a contingent fluke of history that would never be repeated elsewhere, looks unlikely. We find good reasons for having four bases in the code; in the landscape of chemical possibilities, the four bases have certain properties that optimize an information storage molecule and its flexibility and ability to replicate. We also find that the codon table has nonrandomness. Although the exact events and selection pressures that yielded the coding table we observe today are still not fully unraveled, many conditions, from the chemical affinity of amino acids to RNA and the drive to minimize errors, are not mere contingencies, but they emerge from physical and chemical properties, the latter ultimately linked to the physics of atoms.

Like much about biology, before all this knowledge was available, it was almost impossible to predict what the genetic code looked like. No one could have written down the details of DNA in 1950, before the discovery of its structure. This observation leads some people to claim that this is a difference between biology and physics, that physics has laws and equations that are used to make predictions, but biology does not. But this comparison may not be completely fair. We do need to understand the genetic code and its chemistry before we can make predictions, and this knowledge has come only relatively recently in scientific history. In the same way, physicists did need to have some rudimentary knowledge about the behavior of gases under different temperatures and pressures before they could conceive of the ideal gas law, for instance. Indeed, a basic grasp of the genetic code has allowed people to run computer models of error minimization in alternative codes and to predict which ones work and which do not. By running laboratory experiments alongside models using different bases, scientists have been able to explore and predict the efficacy of alternative genetic codes. Synthetic biology, in its quest to make new codes and incorporate them into life, demands better predictive capabilities. The success and accomplishments of synthetic biologists rely on their ability to make predictions about the new compounds or creatures they plan to design.

The sheer complexity of the code compared with, say, a box with helium inside might stymie our ability to use simple equations to predict the code's behavior and puts the code in a very different category for study. The code's complexity, however, does not make the code any less a slave to physical principles, and neither does this history imply that the code is an unlikely product of chance. Although physicists who predict behaviors of gases no doubt face a more constrained problem than do scientists trying to predict the complexity of the genetic apparatus, separating these investigations into two entirely different problems seems misguided. Much about the code opens itself to more simple physical, and thus chemical, principles than was once assumed.

Further down the line from the genetic code to the proteins that it encodes, we find this same apparent lack of chance. Churning out from the RNA, the last step in the code, are long strings of amino acids, proteins that will fold to make the working molecules of life: the enzymes and structural parts to build a cell.

Curious researchers have long wondered whether the number and type of amino acids used in proteins are random, particularly given that in the nonbiological world, there are hundreds of amino acids. Initial attempts to discover if, given a set of random alternatives, evolution would select the twenty amino acids predominantly found in life were inconclusive but pointed tantalizingly to the possibility of nonrandomness. But then in 2011, Gayle Philip and Stephen Freeland published a refined study in the journal *Astrobiology*. They began with the assumption that of all the properties of amino acids essential to protein structure, three are of special importance.

First, the size of amino acids determines how the long amino acid chain, which constitutes a protein, folds and whether it can be properly bundled together into an active molecule. Second, the charge of an amino acid also plays a key role in a protein. Negatively and positively charged amino acids can be attracted to one another and form a bridge that holds a protein together. Many of these bonds dotted through the structure are one of the most important means by which the whole necklace of amino acids can be brought together into a well-defined, ordered structure able to carry out a useful function. Third, repelling water (hydrophobicity) is yet another very useful feature of amino acids. As proteins are dissolved in water or sometimes in membranes where there is no water at all, different affinities for water among the amino acids turns out to be crucial for molding the behavior of whole proteins or parts of them. It alters how they attach to other proteins and whether they have an attraction for regions of the cell that lack water, such as the deep interior of the cell membrane.

Philip and Freeland chose some amino acids and then ran a program that selected a group of them according to whether they had a wide range of sizes, charges, and hydrophobic nature. Their program also chose amino acids that would not only have a wide biochemical range, but also have an even distribution across that range so that their biochemical properties did not overlap too much in one area. This distribution, they assumed, would provide the best tool kit for life since a wide range of properties for the proteins was available. If they are evenly distributed, then life can also choose amino acids that have a high chance of being close to what it would ideally like to use. It is a sort of pick and mix of characteristics like the varied wrenches you might have in a DIY toolbox. You do not want all the wrenches to be very large or small; rather, you want an even distribution, a variety of sizes that have a good chance of including the one you need to undo that bolt on the old door you are trying to remove.

The first set of amino acids Philip and Freeland included in a study of "coverage" (their term for a wide, even distribution of properties) were the amino acids found in the Murchison meteorite. On the assumption that amino acids like these would have been raining down on the Earth when life first emerged, it seemed reasonable to test them first. They tested a selection of amino acids among the fifty found in the meteorite. These fifty included eight actually used in life and another forty-two amino acids that as far as we know are not found in living things. Philip and Freeland ignored some of the branched amino acids (sixteen in total) that were deemed too large and obstructing to be plausibly used by life to make proteins.

What they found was astonishing.

When they compared the twenty amino acids used by life with a million alternative bundles of amino acids randomly chosen from the fifty in the meteorite, the twenty used by life had better coverage and combinations of all three of the key factors than did any other set. The amino acids used by life appeared to be anything but random. Instead, they seemed to be selected by evolution to give a wide range and even distribution of properties that might be useful in proteins—what one might expect from a versatile and flexible tool kit.

Nevertheless, only eight amino acids used by life were found in the meteorite, and the remaining twelve that life uses are derivatives of the first set of eight primitive amino acids. These twelve derivatives were made possible by new synthetic pathways in the cell. So the researchers reran their analysis, again using the fifty primitive meteorite amino acids, but this time searching for only an optimum group of these eight very primitive amino acids. Less than 1 percent of these groups were better than the natural eight used in life, and less than 0.1 percent were better across all three characteristics. Again, the results were uncanny.

In these last calculations, however, maybe we have some enticing evidence of groups of amino acids that might be better than those used in life? In life, we seem to have nonrandomness, but might there be sets even more promising than the chance selection of the eight used by life? We should be cautious asking such questions. As Philip and Freeland themselves recognized, they chose only three features of amino acids, and there may be other factors that decide how useful amino acids are, such as their ability to move around in a protein chain (steric or structural factors).

In their final test, the researchers made an even larger group of amino acids. Fifty came from the meteorite, but they were augmented with the other twelve that life uses. Philip and Freeland also threw into the mix another fourteen amino acids that are made in the cell as intermediate compounds in the synthesis of the twelve made in the cell and used in proteins; these fourteen are not actually encoded in DNA. From this much-expanded set of seventy-six amino acids, taking random groups of twenty amino acids, not a single group out of a million possible alternatives outperformed the natural set.

Philip and Freeland's results are provocative. There is still much we do not know. Which amino acids were in abundance on early Earth to be commandeered by life? Are other properties of amino acids of importance for life in deciding which ones get picked? Doubtless, as knowledge of early Earth and proteins converges, these sorts of studies will be improved. Barring some strange coincidence and bad luck in running the program that has taken us down an egregious blind alley that will one day be corrected, Philip and Freeland's research does strongly suggest that the twenty amino acids predominantly used to construct living things are not random. They have been selected for their collective versatility in providing a range of properties that life can sample to build the huge array of proteins from which the earliest life was assembled.

In more recent years, synthetic biologists, not content with changing the genetic code, have had great success in getting cells to incorporate new types of amino acids into proteins. With modern molecular tools, the hope is that some amino acids not used naturally in life might find use in proteins to make new therapies for diseases. Designer proteins, constructed from sets of amino acids not encountered in natural biochemistry, raise both enormous scientific potential and ethical questions.

When looking at these newfangled creations, we might be tempted to use them as evidence that life is biochemically so flexible that the existing amino acids used by life must be just a fluke, a frozen accident. After all, if some of these new amino acids can accomplish new biochemical tricks, is it not the case that life has failed to tap into these capabilities, because to swap to the new amino acids would mean too great a disturbance to its existing pathways? Given the chance to rerun evolution, maybe life would find these new and exciting biochemical properties from scratch, building different sets of amino acids that would include those now in use by synthetic biologists?

There is, however, a crucial difference between evolution and synthetic biologists. The scientists find particular biochemical properties that might be useful, perhaps to make an effective drug or a new compound of use to an industrial chemist or a pharmacologist. With forethought, the scientists can select an amino acid and incorporate it into the cell to achieve a desired result. Life, however, must select a set of amino acids that can be used across a vast number of proteins, and it must do this to optimize its energy demands. Having ten different sets of twenty amino acids, all of which might do something useful, costs a lot in terms of materials and energy. A cell that replicates sufficiently in the environment to proliferate with fewer energy-demanding pathways will likely be at an advantage. The same argument is true for expanded genetic codes. That we can add letters to the code at will and even make microbes with stable, larger alphabets in the laboratory does not demonstrate that over many millions of years, under exposure to natural environments and competition for food and resources, these expanded genetic codes would confer a long-term advantage to organisms against those with a code containing our familiar four letters.

Philip and Freeland's work illustrates that the pressure on life is more likely to result in a small generic tool kit of amino acids with a wide, even distribution of biochemical properties that maximizes the possible things that life can build with that set. This biochemical evolution differs greatly from the motivation and pressures that channel the energies of a synthetic biologist. The mere possibility that a cell can use an extravagant variety of amino acids to make proteins when coaxed by a scientist to do so tells us little about whether the selection pressures on a whole organism in the environment would preferentially land on those amino acids. The demand on life is rather to select the minimal number of maximally diverse types.

We know that life can take diversions. The unusual amino acid selenocysteine is found in some proteins. The selenium atom within it seems to improve the ability of proteins to deal with antioxidants.

Another strange cousin, pyrrolysine, is an amino acid found in some methane-producing microbes. Both compounds expand the set of amino acids in life to twenty-two, showing that when confronted with the need to expand the repertoire of protein building blocks, driven by some specific biochemical requirement, life can achieve this.

The genetic code, with its number and types of bases; the codon table, which specifies the amino acids the table will code; and even the amino acids themselves are all apparently limited, nonrandom choices. However, perhaps all this does not matter. With just 20 possible amino acids, we have unlimited potential to string together a simply vast number of molecules. Consider a protein with 300 amino acids. Each place in the chain could have 1 of 20 amino acids. When strung together, these 300 places for any of 20 amino acids represent 2×10^{390} different possible combinations! That is enormously more than all the stars in the known universe. Therefore, with just a limited alphabet of amino acids, life has the unconstrained potential to create diversity, quirks, and experiments in design that knows no bounds. Here, surely, at the end of the trail, as the chain of amino acids finally folds up to produce a molecule, we are in the realms of chance. With such diversity, surely now we leave the constricting behavior of physical limits and open up a world of molecules where life in all its variety will be unbounded?

When biochemists first began to explore the mesmerizing variety of proteins from which life is assembled, they seemed confronted by a paralyzing number to grapple with. With 2×10^{390} possible different sequences from a single hypothetical protein chain of just 300 amino acids, how many centuries of biochemical work would be necessary to make sense of all the molecules that exist in the real world? Yet as these molecules were uncoiled, their strings of amino acids read and their folds studied, it became apparent that no matter what the sequence of amino acids, the number of folds or shapes that parts of proteins could adopt was very limited indeed.

Pull proteins apart into their individual units, and you will find that they parade a very meager set of folding arrangements. Helices (termed α-helices) are right-handed helical arrangements of amino acids held together by a hydrogen bond between the hydrogen on an amino group and an oxygen of an amino acid three or four places earlier in the sequence. Another type of fold are the pleated sheets (usually called β-sheets). These are long chains of amino acids held together by hydrogen bonds to make sheetlike arrangements.

These two types of folds can be strung together to make combinations. Many proteins are made up of α and β structures, assembled from helices and sheets that occur in various permutations along the amino acid chain. In some proteins, the two forms occur in strictly alternating forms (α/β). These structures are themselves subclassified into triosephosphate isomerase barrel, sandwich, and roll motifs, which are particular ways in which the helices and sheets can be folded. Never mind the finer points; what we see here is a bounded set of possibilities.

One explanation for this small chorus could be that these folds became locked into life early in its evolution and that, being sufficient to assemble something that is useful, there is no longer a selection pressure to evolve further forms. The analogy is like building a house. You do not go to a builder's store and use every brand of brick available. You select a few that will do the job. Once an ancestral organism made the choice of folds, the rest of life was stuck with it.

Compelling though that argument may seem, much more fundamental laws at work may select for the arrangements of protein folds. Amino acid chains will collapse in such a way as to arrive at a low-energy state. The successive folding steps are driven by thermodynamics to their most stable state. The folds are not independent, and different parts of a protein exert an influence on the folding patterns of other parts. As all these folds collapse into the final product, the protein seeks the most thermodynamically favorable configuration. This means there are only a few solutions. Are the laws of physics disobeyed, or entropy violated, if a strand of disordered amino acids neatly packages up into a ordered machine ready to do some work? Not at all. When amino acids arrange themselves into a protein, water molecules are being forced out of the structure into the surrounding milieu, into the disordered chaos of water molecules in the outside environment. The second law of thermodynamics is not violated in this transition to molecular order.

Here again we see a beautiful synergy between biology and physics, sometimes instead framed as a polarized difference. Some people perceive a conflict between two possible views, between the existence of biological "laws" that drive life to a few simple and predictable solutions and a different, "Darwinian" perspective of evolution, where there is no preordained order, and variation and selection define a vast landscape of possibilities. However, the two viewpoints seem compatible and inseparable. Darwinian evolution, through genetic variation and selection, experiments with a great diversity of forms, but those forms still conform to the laws of physics and are tightly constrained by the universal principles that operate at whatever scale we are observing. With proteins, Darwinian evolution generates many proteins with diverse structures useful for different functions, selected because the processes in which they are embedded are beneficial for survival. However, thermodynamics greatly restricts the number of shapes in which this glut of molecular forms can be assembled.

The study of the genetic code and its translation into the fabric of living things has occupied many people. Some have a fascination for DNA, others for protein, still others for that long-lost Narnia of early Earth, where the first RNA molecules may have leaped into chemical activity and life. A few spread themselves across this landscape of biochemistry. However, over the last few decades, independently across this vista, scientists have apparently removed contingency from much of what was once considered a virtual miracle of machinery. Life was believed to be a system of such extraordinary molecular complexity and yet, in its functioning, a system of such elegant simplicity, it seemed the whole thing must have been chance, an accident, one of many paths that life might have taken. Yet physical and chemical constraints appear to have hammered and forged the code of life in ways that now, with computational methods, even open themselves to comparison with alternative worlds. The fog lifting over this once-unnavigable diorama of molecular forms reveals it instead to be aligned and arranged in patterns more clearly seen.

CHAPTER 8

OF SANDWICHES AND SULFUR

WHEN I UNWRAP ANOTHER sandwich at the café in the building in which I work in Edinburgh, it doesn't usually consciously occur to me that this culinary delight is not a mere sandwich, but a bundled-up package of tasty electrons, subatomic particles that come in lettuce, tomato, and chicken flavor.

Yet, hidden behind its sandwichy guile, this tempting cardboard-encased cocoon of university catering sumptuousness is nothing more than a convenient way of consuming electrons. Throughout the vast diversity of living things, from the smallest bacterium to a blue whale, there is a stunning commonality in how the cells of these creatures get their energy to grow and reproduce. So identical is this process across life, so simple and in which such basic principles inhere, that it is easy to imagine that life anywhere across the universe might get its energy in the same way. It is in this machinery that we pursue our adventure into the structure of life, exploring its basis in physical processes. From the codes and molecules that build life, we turn to another vital piece of life's molecular machinery—how it gathers energy from its environment to grow and reproduce, the process that powers the biosphere.

In the 1960s, a brilliant scientist, Peter Mitchell, pondered the basic mechanisms of how life gathers energy from the environment. He wondered about this because he knew the question was important. The second law of thermodynamics that drives the universe inexorably toward disorder or increasing entropy is a fact of the universe; hence, it is a law, and life must conform to it. Constructing complex machines that can grow and reproduce requires energy to maintain this order against the ever-present second law, which would like to dismantle those machines and dissipate their energy and the components of their molecules into the void. Fathoming how life gets energy from its environment is therefore not merely of interest to apprehending how it interacts with the world around it, but it is fundamental to knowing how it operates within the constraints of the laws of the universe; few are as basic as the second law. In the temporary oasis that is our Earth, receiving energy from the Sun and producing its own from the primordial heat within the planet, how does life gather its energy to garner such organizational complexity and then to spread tenaciously across and within our planet?

Mitchell's biochemical insights eventually won him the Nobel Prize in 1978. Like many findings that reshape our worldview, they seem obvious with the benefit of hindsight, but putting the bits together in a way that, to later generations, would appear to be common sense required a stroke of creative genius. The result was yet another foundation stone in our understanding of biology—an understanding that speaks of general applicability, another basic piece of machinery whose roots in physics suggest the potential similarity of life across the cosmos, if indeed it has evolved elsewhere.

But back to the sandwiches. Once consumed, where do those sandwiches go? They are disassembled in our bodies into their constituent sugars, proteins, and fats. Some of that material is burned in oxygen to release energy, while some of it may be indigestible. Now in secondary school, you may well remember being tirelessly forced to write down the reaction for aerobic respiration, and dull it was. But bear with me, because one thing they never taught you at school was just how beautiful this process is.

$$C_6H_{12}O_6 + 6O_2 \rightarrow 6CO_2 + 6H_2O + energy$$

There on the left we have a complex chemical formula, $C_6H_{12}O_6$, which is the formula for glucose, a sugar, but it could be any complex carbon compound, from the ingredients of sandwiches to salami. We add this carbon compound to oxygen in the air we breathe, and when we do, these two compounds make energy, with carbon dioxide (CO_2) and water (H_2O) as waste products on the right-hand side of the equation.

The organic material, the sugar shown above, contains electrons, each in a fuzzy orbit within its atom. The electrons contain energy, and it is this energy that life seizes from the reaction. But how does it do that?

All atoms in the universe have varying degrees to which they will give up their electrons. Many of these atoms are electron donors, relinquishing electrons with glee, but others, electron acceptors, prefer instead to take them up. Whether an atom donates or accepts electrons is influenced by a concatenation of things, from pressure and temperature to acidity, but we need not concern ourselves with this detail here. Crucial to your lunchtime hunger is that many organic materials, including the components of your sandwiches, are good electron donors.

Sitting on cell membranes or the membranes of the organelles within them, for example the mitochondria in your cells, there are molecules that will bind to compounds with electrons ready to be grasped. The electrons have now begun the first stage of their journey. Through an act reminiscent of a relay race, the electrons are passed from the broken-down products of your sandwiches to the cell.

Sitting next to the molecule that has just grabbed the electron is yet another molecule that would like the electron even more, and so the particle begins its traverse through the cell membrane, leaping from one molecule to the next. The relay race is under way. Eventually, the electron will get to the end of the race, and what then? There sits an electron acceptor that would like to grab the electron and carry it off. In our bodies, that is the job of oxygen. The electron acceptor is crucial because if you do not take the electrons away, they will get clogged up in the cell and we quickly get an electron traffic jam. The transfer process we have just been talking about will grind to a halt. And now you can understand why breathing is important: getting that oxygen into your body to stop your energy machinery from overloading is rather crucial.

As the electrons shunt through the membrane, their energy is released and we now must do something with that energy: we must gather it up. Mitchell creatively figured out how this was done. As each molecule gains that minuscule amount of energy from the electron as it passes by, the molecule uses the energy to move yet another subatomic particle, a proton, from the inside of the membrane to the outside.

What we now have is a proton gradient: more protons are on the outside of the membrane than in the inside. Those protons, through osmosis, now want to move back into the cell to equalize the gradient.

Place a raisin in a cup of tap water, and the fruit will expand as the salty and sugary interior sucks up water from its surroundings. This is osmosis at work. The water will move in a direction where it can end up with equal concentration inside and outside the raisin. In the same way, as we now have more protons on the outside of the membrane than on the inside, the interior will slop them up until the concentration on both sides of the cell membrane is the same.

The protons sitting outside the membrane have a Janus-faced quality: not only are they at a higher concentration, but they also have a positive charge to them, each one written as H^+. It is both this higher concentration of charge on the outside of the membrane, designated $\Delta\Psi$, and the higher concentration of actual protons themselves, written as ΔpH, that creates this powerful gradient. We call this gradient the rather dynamic-sounding *proton motive force*, Δp, and the equation to work it out is below:

$$\Delta p = \Delta\Psi - (2.3RT/F)(\Delta pH)$$

where R is the universal gas constant (8.314 J/mol/K), T is the temperature of our cell, and F is the Faraday constant (96.48 kJ/V). A typical value for the proton motive force is about 150 to 200 millivolts.

So the protons have a tendency to be drawn into the cell. This they now do, but not by randomly diffusing through the membrane anywhere they please, since the membrane is generally impermeable to them. They flow back through a complex little machine called *adenosine triphosphate* (ATP) synthase, whose job it is to make the energy-trapping ATP.

As the protons flow back through ATP synthase, they cause the pieces of the molecular machine to rotate. This incredible contraption is made of no less than six different protein units. The changing shape of ATP synthase as it goes on its ratchet roundabout brings phosphate groups into alignment with adenosine diphosphate (ADP) and forces them together to make ATP. These new phosphate bonds have now trapped the energy of the electron transport chain.

The molecule so produced, ATP, can be transported around the cell, and its phosphate groups can be broken off to release their energy anywhere it is needed, for making new proteins, repairing old ones, and eventually making new cells. If you think this is a trifling process, your body, in all its cells, produces about 1.4×10^{21} molecules of ATP per second! It has cost you about 2.5×10^{24} ion molecules of ATP just to read this chapter.

Ponder this whole process. There is certainly complexity in all this: different molecules for collecting electrons, the machinery for making ATP, and even the ATP itself, which is a small molecule, but one that nevertheless has a subtlety and intricacy in how phosphate bonds are used to trap energy.

However, at the core of this process is an incredible simplicity. Here we have readily accessible subatomic particles, electrons, with some energy to give away being used to produce a gradient of another subatomic particle, protons. This gradient is then employed, by the rather basic principle of osmosis, to do work in rotating a miniature machine that produces a molecule that effectively stores that energy for release anywhere it is useful. This is a mechanism of alluring modesty.

I have heard it said that this system of gathering energy is highly idiosyncratic. With the benefit of hindsight, it looks simple. However, would life elsewhere really use such a system? To put the question slightly differently, give an engineer a pad of paper and a pen, and ask them to devise a system for harvesting energy from the environment. Would they come up with the same thing?

For the unconvinced, I would merely observe that engineers have done almost the same thing. Hydroelectric power, used in over 150 countries around the world, is made by damming water high in a lake and then allowing it to flow down a mountainside, using the kinetic energy of the water as it rushes down a hill to create rotary motion in a turbine that generates electricity. The details are different in the cell, but the basic idea is the same. Life pumps protons into a reservoir outside a membrane. The membrane acts as a dam to trap the protons outside. Then it uses the gradient of osmosis to produce rotary motion in the ATP synthase as those protons rush back through our molecular turbine. Rather than producing electricity, the minuscule turbine is instead used to store energy in ATP. However, even ATP is analogous to the storage of electric power in batteries for use at a later time or place. In the cell, we do not have hydroelectric power, but we have proton electric power.

Let us imagine the implausible scenario that the field of biochemistry was never developed, but that for some obscure reason, we knew that cells have impermeable membranes. Now give an engineer a pad of paper and a pen, and ask them to devise a system of gathering energy from atoms. I see no reason why they would not, by thinking about hydroelectric power stations, conceive of a system in which electrons were used to make some sort of gradient of ions with microscopic pumps. That gradient would itself be used to channel those ions back into a cell through a rotary device to make electricity or chemical compounds that contain energy.

The chemiosmosis theorem, Mitchell's brainchild, has a certain logic to it, but one might still ask, why the proton gradient? Each movement of an electron from one protein to another releases a tiny amount of energy. We want to gather it all up. Making a gradient by pumping protons is clever because each transfer of an electron contributes to a proton gradient outside the membrane, and the gradient is the accumulated product of many of these electron movements. We end up with a relatively large accumulation of protons, like our water in a mountaintop reservoir. These collected protons can now be channeled through a single machine to capture the energy.

The quintessential question that we might ask again is whether there is any room for chance and contingency in all this, any room for serendipitous attributes of historical quirks. Or is the architecture of this process locked into an unyielding pattern?

We already know that within this energy-making machinery, there is room for flexibility in the detail. Some microbes can use sodium ions (Na^+) instead of protons to generate their gradient. In an ingenious study, a group from Germany used chemicals that punch holes in the membranes of bacteria to show that when the membranes had holes that would allow protons to uncontrollably leak into the bacterium *Acetobacterium woodii*, the organism's ability to use the chemical caffeate in its electron transport chain was unaffected. However, when the holes in the membrane allowed sodium ions to leak through, the organism's ability to make ATP was destroyed. Here is a microbe that has eschewed protons for sodium ions, a microbe that illustrates the possibility of modifications to this ingenious apparatus. Nevertheless, protons are the most common way to make a gradient across the membrane for all the major domains of life, and the use of proton gradients may be no coincidence, but rather deeply embedded in the origin of life itself.

This machinery for extracting energy has an even more remarkable versatility to it. And this is where things turn extraordinary.

Life does not necessarily need sandwiches and oxygen. We can use different electron donors and acceptors and make a life form that can grow using a whole variety of other things to be found lying around in the universe. You and I need oxygen to breathe, but oxygen is not the only chemical that can carry off electrons from the cell. Many microbes instead use iron or sulfur compounds that, like oxygen, seize electrons. The result of this switch is the anaerobes, microbes that can live without oxygen. Deep underground or in a muddy, fetid pool, these microbes go about their lives without a hint of the gas, growing in rocks, in bogs, or deep in the sulfurous pools of volcanoes. These are creatures breathing bounteous elements such as iron, sulfur, and their respective compounds. This simple swap of

electron acceptors opens up a new landscape of habitats and environments for life where humans cannot go.

The potential for life does not end there. Not only can we forsake
oxygen for another electron acceptor, but we may also select other
electron donors—away with sandwiches! Swap them for hydrogen,
and we now have a microbe that can instead use hydrogen gas from
deep underground as a source of food. These so-called chemolithotrophs, literally, chemical rock eaters, have many advantages over us.
Freed of the need for organic matter, they can now live a life essentially independent of the rest of the biosphere, even underground in
the absence of light.

The limitation of using organic matter, including the sandwiches,
as a food source is that its components come from other microbes,
plants, and animals. This interdependency between different types
of life is the basis of food webs that make up much of the ecology of
our planet; herbivores eat plants and carnivores eat the herbivores
and other carnivores, nothing more than a complex web of electrons
moving around from one life form to another. However, by feeding
off hydrogen, microbes are feasting on the raw materials of planets. No group of people is more fascinated by the chemolithotrophs
than are astrobiologists, as they wonder whether these metabolisms
would allow life to live deep underground in habitable regions of
other planets.

By mixing and matching electron donors and acceptors, we
can make energy from a wide range of chemicals on offer. The methanogens, microbes that make methane gas, use hydrogen gas as
their electron donor and carbon dioxide gas as the electron acceptor.
Hydrogen gas can be ancient, trapped in a planet during the planet's formation, or the gas is produced when certain minerals react
with water in the process of serpentinization. Creeping through rock
fractures, dissolved in water, hydrogen can drive whole ecosystems.
Microbial communities that use hydrogen as their main source of
electron donors inhabit many of the boiling volcanic pools in Yellowstone National Park, the hydrogen produced deep underground in
the magma-heated depths of this dormant supervolcano.

The carbon dioxide that methanogens use to carry off those spent electrons is in no short supply, either. At a tiny fraction of the atmospheric composition, about 400 parts per million, the gas is still abundant enough to be used by the microbes as an electron acceptor, although it can be even more concentrated deep in the Earth.

The methanogens spark the enthusiasm of astrobiologists. Methane has been detected on Mars and in the plumes of Enceladus, one of Saturn's moons. Is it the by-product of life? Well, there are ways in which methane can be produced without living things, so the mere presence of the gas is not an unequivocal sign of life. Methane can be made deep underground where gases react at high temperatures, and it can be trapped at low temperatures in ices, so-called clathrates, later to be released if the ices are warmed by volcanic activity. Nevertheless, the very controversy about the origins of methane drives astrobiology missions to find out whether its presence on other planets could be a sign of biology. Behind this quest to test the hypothesis that these faraway places host extraterrestrial life, the motivating driver for all this research is our knowledge of the amazing capabilities of the energy-producing machinery of life and the possibilities it hails.

The electron transport chain is like a sort of modular energy system. The core molecules involved are very similar across different forms of life, built up from cytochromes, other proteins, and quinones that contain within them arrangements of iron and sulfur atoms particularly good at transporting electrons. At each end of the chain are the molecules that trap different electron donors and acceptors depending on where a creature lives and what is available to eat. Cells are by no means limited to one choice, either. They can bolt on new electron donors or acceptors depending on what is available around them. Like a hungry diner in a buffet switching to pasta when the pizza has run out, microbes can shift from iron to sulfur compounds and back again as the available energy sources change, giving them incredible versatility in the places they can subsist.

One most astounding discovery of recent years is that microbes can even use free electrons, isolated electrons not associated with anything. Place an electrode into the sediment, and microbes will attach to the electrode, extracting the electrons directly from it to power their electron chains. A surprising number of microbes— *Halomonas*, *Marinobacter*, and others—have this ability. While the discovery that microbes can use electrons directly to make energy seems extraordinary, it perhaps should not surprise us. Many compounds I have been talking about, including those in your sandwiches, are merely containers for electrons. When free electrons are available, why not cut out the intermediary and take them directly?

The consequences of the energetic versatility of these electron transfer chains cannot be underestimated. Each year, about 160 million tons of nitrogen gas from the atmosphere are taken up by so-called nitrogen-fixing bacteria and turned into ammonia, nitrites, and nitrates, more biologically available forms of nitrogen that feed the rest of the biosphere. Microbes that use these nitrogen compounds in electron transport chains to gather energy carry out all these transformations of nitrogen, from ammonia to nitrites and nitrates and back out again to the atmosphere as nitrogen gas.

The same too with sulfur compounds. Elemental sulfur, thiosulfates, sulfates, and sulfides are all shunted around between different microbes and transformed back and forth, one into another, in global biogeochemical cycles that churn and turn elements and compounds through the Earth's crust, providing them to the rest of life, including you and me.

Probably one of the most interesting and profound discoveries in biology over the past few decades—a consequence of our insight into life's energy-extracting machine—is that almost every electron donor-acceptor pair that theoretically might provide some energy for life has been found in nature. Any combination of two elements or compounds for which it is thermodynamically favorable for an electron to move from one entity to the other and give off energy is fair game.

In a now seminal paper published in 1977, Engelbert Broda, an Austrian theoretical chemist, using some simple energetic and thermodynamic intuition, predicted the existence of microbes in the wild that hitherto had not been discovered. One of them was a bacterium that would use ammonia as the electron donor and nitrite as the electron acceptor. This anaerobic ammonia oxidation or *anammox bacterium*, as it was eventually called, was finally found in the 1990s. The process it drives turns out to be enormously important in the marine environment, accounting for about 50 percent of all the nitrogen gas produced in the oceans.

Here we have an example of how knowing about the physics of life allowed for a prediction of a life form subsequently discovered. The view that physics is underpinned by laws that allow predictions and that biology is so varied that it lacks the predictive rigor of physics falls away. We see in the energetics of life simple thermodynamics embedded within the molecular machines of energy production. These basic principles allow us to predict the energy-gathering capacities of living things equally as well as apparently more simple energetic systems.

Some of the energy gathering that microbes carry out has found very practical application in some surprising places. Instead of using oxygen as an electron acceptor, some microbes can use uranium, perhaps to be found in a contaminated nuclear waste site. The microbes, by using uranium in their electron pathways, alter the chemical state of the element. This new form of the element is less easily dissolved in water and is consequently less likely to leach its way into the water supply. Using microbes to change the state of hazardous chemicals in the environment into forms less likely to cause harm or be a public health risk is the ingenious process of *bioremediation*. Microbial energy gathering has now gone beyond pure academic knowledge and entered into the service of humanity in solving some of our emerging and urgent environmental problems.

Before we lose sight of our purpose, I want to return to what all this means for evolution, for life and its possibilities.

Combining sandwiches with oxygen produces a lot more energy, typically about ten times more, than do many reactions that use chemicals like iron or sulfur. Anaerobic lifestyles are quite energy poor and the microbes feeding off iron in the rocks or chomping on hydrogen deep underground are living life on the edge, the thermodynamic edge. If you want to run a brain (which in humans requires about twenty-five watts), run, jump, fly, and operate a body with many trillions of cells, you need a lot of energy. Those energy-yielding reactions that happen without oxygen are generally just too feeble to be used by most animals. In anaerobic habitats, life is limited by energy, yet another boundary set by physical processes.

It is apparently no coincidence, then, that animal life on Earth emerged when oxygen levels in the Earth's atmosphere increased to approximately 10 percent, the threshold at which the energy from aerobic respiration may have supported much more complex life. Aerobic respiration could have occurred at lower concentrations of oxygen than this, but life would have been denied the large-scale complexity we associate with animals. It took a revolution in energy acquisition to allow for the emergence of the biosphere with which we are all familiar.

However, why did the concentrations of oxygen in the Earth's atmosphere rise to allow for this dramatic increase in energy availability? We know that the oxygen gas in the Earth's atmosphere came from photosynthesis. Cyanobacteria, pervasive green microbes that occupy the oceans, lakes, and rivers, figured out how to make energy from sunlight by splitting water molecules to release their electrons. Sunlight is used to energize the electrons that eventually end up running through our trusty electron transport chains to produce ATP. The splitting of water molecules for energy was a revolution because until then, life that used sunlight as a source of energy was confined to using chemicals like hydrogen and iron as their source of electrons. By switching to water, a most abundant and widespread resource, the oxygen-producing photosynthesizers conquered the Earth's landmasses and waterways, presaging a huge production of oxygen gas.

Unfortunately, the newly available oxygen was not immediately free to build up in the atmosphere. Because copious quantities of gases such as methane and hydrogen like to react with oxygen, the concentration of these other gases had to be lowered before the concentration of oxygen could rise. The chemical evidence locked in ancient rocks suggests that this increase in oxygen happened about 2.4 billion years ago in the Great Oxidation Event and again about 750 million years ago; the second increase produced concentrations high enough to allow for animals. There has been no event of greater consequence for life than the rise of oxygen. This chemical change in the atmosphere is thought to be linked not only to the rise of animal life but also, by implication, to the rise of intelligence.

So animals need oxygen, the electron acceptor that releases enough energy for a monkey to swing and jump through the rain forest, a dog to run and roll in the Meadows, and a human brain to think. But is there really no other way that animal life, let alone intelligence, could gather enough energy to evolve on a planet?

At the end of the astrobiology course I teach at the University of Edinburgh, I finish with a lecture designed as much to educate as to entertain my long-suffering students. I walk into the lecture theater, announce that I am off to get a coffee and that the students will soon be greeted by a visiting lecturer. I return dressed in a full-body lizard-man outfit and face mask to deliver my lecture "Is there life on Naknar 3?"

The lecture begins with a description of a distant extrasolar planet we have discovered. It is large, has oxygen in the atmosphere, and apparently has a moon. It becomes rapidly apparent to the students that the planet of which the visitor speaks is the Earth. The lecture weaves an internally self-consistent story about why this distant planet could not possibly support life. Throughout my discourse, I am forced to break off the lecture to munch on some sugar cubes, which the audience soon discovers is a supply of gypsum, or calcium sulfate. You see, I am a sulfate-reducing anaerobic alien that eats organic carbon, but instead of burning this in oxygen, I use sulfate as the electron acceptor. As this mode of energy production yields about ten times less energy than does aerobic respiration, I must constantly interrupt the lecture to snack.

The high concentrations of oxygen on distant Naknar 3 make life unlikely there because living matter would combust in this gas. And besides, oxygen produces dangerous free radicals very damaging to carbon-based chemistry. To add to our meager assessment of this world, an additional theory is that the surface of this planet is made up of giant sheets of rock that move around, destroying the ancient sulfate mounds that provide food for life and are necessary for the rise of intelligence. Oxygen and moving land (plate tectonics) would conspire to make this place a poor location for life, at least complex multicellular life.

The educational objective of the lecture is to get students to question whether the view we have of habitability and the clement conditions for life on Earth are limited by our own knowledge or whether our planet is truly a universal template for all life in the universe. By producing an elaborate fifty-minute story where all the bits seem to cross-check and yet lead to a conclusion that Earth is uninhabitable, I want to challenge them to think about whether our view of the evolution of life on Earth is simply a giant *Just So* story we have assembled into a jigsaw about something that was serendipitous, a chance result of one specific evolutionary route.

My own view, as I am sure you have now guessed, is that life on Earth tells us something fundamental and universal. There are many flaws I can find in my own elaboration of the unlikelihood of life on Naknar 3 and the idea of the sulfate-reducing intelligence. Being a sulfate-reducing intelligence really would require almost constantly eating sulfate to generate enough energy to power a brain or a body that can walk. However, besides this inconvenience, the problem with sulfate reduction is that it requires geologically abundant and readily available sources of sulfate to allow for the emergence of multicellular life. I cannot rule out that on some distant world where there is a vast quantity of sulfate piled up in ancient mounds, there might just be scope for herds of sulfate-reducing pigs gnawing away at gypsum mounds, grunting and snorting at their intelligent alien handlers who return to their homes to eat gypsum pie. For now, the sulfate-reducing intelligence must remain speculation, and the low amounts of energy to be gathered through this metabolism make the squealing gypsum-eating pigs unlikely. Regardless, even if the squealing pigs do exist, they are still using electron transport to harvest energy, conforming to the principles we have already identified.

Even on our own planet, though, we do know of complex animal life that tentatively explores unconventional sources of electrons for energy. Giant tubeworms, *Riftia pachyptila*, inhabit hydrothermal vents in the deep oceans, where boiling hot water bubbles up from the Earth's crust, often spewing black sulfide-rich minerals into the ocean. These worms contain within them consortia of bacteria that use hydrogen sulfide (rotten egg gas) from the vents and oxidize it using oxygen gas dissolved in the water. There is enough energy there to produce worms over two meters long and about four centimeters in diameter, the strange white cylinders with their dark-red heads swaying and shimmering in the warm water venting from the rocky chimneys that surround them.

At these vents, the concentrations of hydrogen sulfide gas, which are usually unmeasurable in the atmosphere, are so high that the bacteria in the worms' guts can forage enough to use it as an electron donor to make energy and thus synthesize organic compounds, which the worms use to grow to their gargantuan size. Although the electron transport chain still lies at the heart of this extraordinary symbiosis, these bizarre, twisting tubes show us that sometimes on Earth, where rare chemicals and gases reach unusual concentrations, animal life can take on strange alliances to sustain itself by exploring different forms of energy.

These endearing worms show that evolutionary possibilities can be multiplied by the enormous variation in the geochemical availability of the raw materials of life. So far, I have discussed how physical principles greatly constrain the products of evolution within some very narrow bounds. One way in which the scope of evolutionary possibility can be changed within the strictures of physics is by changing the geochemical availability of things needed to drive innovations. Even without wild science-fiction stories about sulfate-reducing lecturers on our own world, we have abundant evidence, not least in the deep-ocean worms, for how geochemical availability can redirect the course and products of evolution. The worms show us how life is bounded by physics all along the evolutionary path, but perturbations in the geology of planets can open new evolutionary vistas, particularly when it comes to the chemicals that can drive energy acquisition.

Nevertheless, the rise of oxygen in the Earth's atmosphere, itself a result of life, made aerobic respiration more widely available. Thus, the rise of oxygen enabled life to access vast quantities of energy, opening the way to multicellular life and intelligence once the earliest animal cells tapped into these opportunities. More energy begets bigger and more-complex machines, triggering the emergence of a biosphere that reached beyond the simplicity of single-celled life.

Woven throughout this evolutionary story, from the earliest microbes to the rise of animals, are electron transfer chains. More than this, the energy available from different electron donors and acceptors, the energy released from these chains, explains much about the transitions that have occurred throughout evolutionary history. Early life soldiered on, extracting energy from rocks, gases, and organic materials. Then concentrations of oxygen gas rose, driven by the physics of splitting apart water and the use of electron chains to capture energy in sunlight. From this momentous innovation, the rise of animals was linked to the thermodynamics of how much energy is released when electrons are transferred from organic matter to this newfound oxygen. Microbes and monkeys roam the Earth, powered by energy released from movements of a subatomic particle, the electron.

It is appropriate at this stage to wonder whether there are any other energy-yielding reactions that life might use as alternatives to electron transfer chains. Perhaps other forms of energy could get around the problem of the limited amounts available in oxygen-free environments. Can we escape oxygen altogether yet gather large amounts of energy to drive a complex biosphere? To say that life is constrained by physics is not to say that it is limited to only one method of collecting energy. As with humans, from wind power to nuclear power, perhaps a self-replicating, evolving system has more than one card up its sleeve.

Certainly, we know that some cells can get away without using an electron transport chain. Instead of all that rigmarole with membranes and electron donors and acceptors, an alternative approach is to just take a phosphorus-containing chemical group going spare on a molecule and tag it directly onto ADP to make that energy-storing molecule ATP. This process is at the heart of fermentation, an extraordinarily versatile set of metabolic pathways that permeate our lives. Fermentation turns sugars into acids, which we use to pickle vegetables, or you can use it to make alcohols, the basis of beer, wine, and much else besides. This same process goes on in your own body as sugar transforms into lactic acid, giving you cramps if you do some sudden exercise and you cannot get enough oxygen to the muscles. Although this process does not use an electron transport chain per se, it still involves chemical reactions that are essentially electrons moving around. And despite the relative simplicity of fermentation, it produces only about a tenth of the energy yield of the electron transport chains. The low energy yield explains why your body prefers to get oxygen and not cramps and why many microbes, when offered the chance, switch to aerobic respiration over fermentation.

However, perhaps we are still not being imaginative enough. What about something much more radical? One line of thinking might be to consider whether there are other particles in an atom, aside from electrons, from which we might get energy. Besides the electrons, atoms have energy bound up in their nucleus. Is the nucleus a possible source of energy for life?

Gathering energy from nuclear fission, the decay of the nucleus in some unstable elements such as uranium, could be one way for life to ramp up its options from the rather paltry pickings available in electrons. Unfortunately, the atomic nucleus is very difficult to control.

We could generate a vast amount of energy by producing a nuclear chain reaction like in a nuclear reactor, but such a reaction requires a lot of uranium or a similar fissile element. There are seldom piles of such an element just lying around in the environment. Aside from using technological ingenuity as we do in nuclear reactors, we would have difficulty seeing how life forms without technology could use this energy and, if they did, how they would control a nuclear reaction without essentially vaporizing in a meltdown or an explosion.

Another way life could tap into nuclear fission is to make use of the by-product of fission—ionizing radiation. As unstable elements such as uranium and thorium decay, so do they release high-energy emissions in the form of alpha, beta, and gamma radiation. Could this radiation power life?

We already know that ionizing radiation has enough energy to split water. The radiation produced from the decay of fissile elements in the planet's crust blasts apart water molecules to make hydrogen. As we have seen, hydrogen can be an electron donor to make energy. Nevertheless, even here, the electron transport chain is at work, taking the products of ionizing radiation and shuttling them through the cell. Nuclear fission is merely an add-on precursor for making electron-containing hydrogen, food for life.

There are even more bizarre environments where we might look for evidence of life that uses ionizing radiation. Human society releases this radiation, unfortunately sometimes unintentionally. One of the more remarkable claims to have emerged from devastated nuclear disaster sites concerns fungi living near Ukraine's Chernobyl nuclear reactor, which blew its top in 1986. Researchers in Russia carried out laboratory experiments using fungi exposed to intense radiation there. The black melanin pigment the fungi contain became better at electron transfer reactions when it was irradiated. Fungi that contained the pigment were more metabolically active, raising the extraordinary possibility that these fungi, and other organisms like them, may even benefit from the radiation streaming out from decaying and devastated nuclear reactor cores. Nevertheless, within this postapocalyptic finding, electron transfer reactions still reign supreme. Indeed, it is very difficult to think of a way that highly energetic radiation from nuclear fission, with a tendency to destroy compounds, can be harnessed by life other than by breaking down molecules into bite-sized pieces that can be fed into the more sedate and kindly electron transport chains, or by changing the electron transfer properties of molecules.

As the fungi at Chernobyl suggest, energy can be harvested as atomic nuclei fall apart in fission reactions, but alternatively, energy is also given up when some atoms are smashed together. Unfortunately, unleashing energy in nuclear fusion to power life seems even less straightforward than fission. Vast amounts of untamed energy can be freed when nuclei of atoms are fused; nuclear fusion is the basis of energy production in the Sun. However, the conditions required to get the reaction going are extreme: temperatures of many millions of degrees Celsius must be reached to coax nuclei to bind. Even brown dwarfs, planets tens of times larger than Jupiter, cannot reach temperatures in their cores to ignite fusion reactions. Efforts to do this using technology require elaborate nuclear fusion reactors to contain plasmas at these unconventional temperatures. Nuclear fusion seems an unlikely source of energy for life (other than, of course, the bright light that is produced from a star's nuclear fusion reactions and that here on Earth is used in photosynthesis). For now, then, gathering the energy of the atomic nucleus directly, bypassing electrons altogether, seems unlikely.

Probing around in the atomic structure to find new sources of energy is one way to seek alternatives. Another way is to think about physical processes that might yield some accessible energy. In an elegant stroll through some imaginative alternatives, astrobiologists Dirk Schulze-Makuch and Louis Irwin proposed some alternative ways of subsisting. Kinetic energy, the energy of movement, for example, in tides or currents in oceans, might be harnessed using small hairs that wave around in the water like the little hairs on the surface of some protozoa. The bending of the hairs in the water currents would trigger the opening of channels that drive ion movement and, like the electron transport pathways, the capturing of energy.

Perhaps thermal energy might be used to take advantage of the huge heat gradients in hydrothermal vents, where temperatures drop from hundreds of degrees to just a few degrees above freezing as the erupting fluids from the crust come into contact with the oceans. Some forms of life might harness magnetic fields to separate ions and, by changing their alignment with a magnetic field, might use the resulting movement in ions to drive energy harvesting. Schulze-Makuch and Irwin also considered osmotic gradients, pressure gradients, and gravitation as ways to drive the movement of ions and molecules and so collect energy to do work.

Before anyone scoffs at these ideas, remember that the basis of the chemiosmosis mechanism is a whirring, rotating machine, ATP synthase. Ultimately, the whole purpose of electron transport is to build up a proton gradient that makes ATP synthase go round and round when the protons flow through it. The mechanical principle is not different from steam being used to turn a turbine; the only difference is that from the turbine, we make electricity, whereas in the cell, we use the changing shape of ATP synthase as it rotates to snap ADP and phosphate groups together to make ATP.

The cell does not care what makes that ATP synthase go on its merry-go-round. With an open mind, we can imagine life separating its rotating ATP synthase from electron transport and proton gradients and perhaps directly tapping into the movement of ions driven by other gradients: gravity, pressure, heat, and magnetic fields.

However, as Schulze-Makuch and Irwin recognized, many of these radical energy sources have problems. Gravitational energy is really too small to move things around at the microbial scale. The same is true with pressure gradients. The difference in pressure between the surface of the Earth and the deep interior of the planet is enormous, but at the microbial scale, it is paltry (0.01-pascal difference between one end of a one-micron-long bacterium and the other if it is pointed downward). Can life use this differential to do anything useful? An apparatus that would use such a small gradient seems unlikely. Magnetic fields on present-day Earth are so small that it is difficult to see how they could be used to make energy, although some microbes and animals do detect these fields and use them to navigate. On other planets, stronger magnetic fields might offer greater potential for energy generation.

Other problems confounding these ideas are the special environmental conditions for them. Large thermal gradients are common around deep-ocean hydrothermal vents, in deeply heated rocks, and on the surface of the Earth warmed by the Sun, but they are not ubiquitous. These gradients would probably have to be highly stable, intense, and reliable for microbial communities to develop and be sustained over long periods. Osmotic gradients using salts and ions also depend on a wide diversity and persistence of these gradients.

No doubt, the physics of alternative methods of gathering energy in cells is well worthy of investigation. Their demonstrated plausibility, even discovery, in the environment would expand our appreciation of how life can, within the constraints of physics, exploit free energy in the environment. We could better say whether chemiosmosis results from strange idiosyncrasies in the Earth's evolutionary experiment or if it reflects a much more fundamental and predictable path in the physics of energy acquisition by life. However, even assuming that we have not found these other pathways because we have not looked hard enough (and there are many unassigned genes in the DNA of living things), none have jumped out at microbiologists and molecular biologists.

With these other possibilities in mind, do we consider Mitchell's chemiosmosis system pure chance at work, an example of where predictions from physics must be left behind to allow for the vagaries of randomness? Was it a contingent invention, stumbled across and then locked into life as a historical accident? Would other methods of energy acquisition dominate on other planets? I do not think so. As much of this chapter has shown, the success of the electron transfer chain is the ability to tap into a wide diversity of electron donors and acceptors and even pure electrons themselves. Living things that can use electrons this way have at their disposal a vast array of elements and compounds from which they might collect energy anywhere, from the surface of a planet down into its interior. On the surface, electron transport can be used to harness the tremendous energy available in photosynthesis. As environmental conditions change, life forms that can plug and play different electron donors and acceptors have a huge advantage in moving into new habitats where resources are unexploited. Someone else has eaten the hydrogen? That's no problem; I'll eat iron. The sulfate has all gone? No worry; I'll eat nitrates. The ability to strip electrons off the raw materials of planets or the by-products of other living things to make energy has a spectacular flexibility that few other energy sources have.

Certainly, we can imagine bacteria that eat ionizing radiation or long, tubular life forms that use thermal gradients across a hydrothermal vent. Nevertheless, we cannot help but conclude that on any planet, access to the multifarious forms of electrons in anything—from water to uranium—would give any life form an advantage. The ability to tap into this energy would give life a great step forward in colonizing the diverse environments that any geologically active planet produces. We must always be careful about confirmation bias, but I think there is a logic to, a reason for, the success of electron transfer as the core energy machinery of life on the planet, and I suspect it would be the case elsewhere.

Like life on the macroscopic scale, we can imagine variation. We know that sodium ions have been used instead of protons for making a gradient across a cell membrane; perhaps other ions might be used. Maybe proteins in the electron transfer chain might have different structures from the ones we know on Earth, mirroring the huge diversity of these proteins even on our planet, but still containing iron, sulfur, or other elements suitable for shuttling electrons. These variations are analogous to the different colors and modifications in shapes of animals on Earth at the scale more familiar to you and me. But at the core, at the bottom of all this machinery, is the manipulation of a subatomic particle, the electron, to gather free energy from the universe. Nothing reflects so beautifully the potential universality of living systems, their link to the most basic particles and physical principles in the cosmos, and that physics and biology are utterly inseparable.

CHAPTER 9

WATER, THE LIQUID OF LIFE

SAMUEL TAYLOR COLERIDGE WAS not an astrobiologist, but his mariner's observation that there was "water, water, everywhere" was about as important an observation as you could make about this most fundamental requirement for life.

There is a lot of water on Earth, about 1.4 billion cubic kilometers, or in more mundane terms but even grander numbers, about 560 trillion Olympic-sized swimming pools. Only about 0.007 percent of this water, freshwater, is actually used by you and me. The rest of it is in seawater, estuaries, marshes, swamps, and deep underground, inaccessible to humans but available to much of the rest of the biosphere, such as the microbes.

Carrying out the chemical reactions for life in a liquid makes sense. In a fluid, molecules can be brought together close enough to carry out reactions. Importantly, many millions of molecules can move around and meet in many combinations, aligning to do chemistry and to drive the complexity of pathways in living things. Such interaction is usually difficult to achieve in a diffuse gas cloud or in a solid. In a solid, molecules and atoms are generally rigid and cannot move around easily. In gases, they are often too far apart, in other words, too diffuse.

We might argue, with a little imagination, that in a gas, molecules would just react slowly and meet each other infrequently. The quixotic intelligent interstellar cloud, the product of imaginative science fiction, for instance, is perhaps just a bit cumbersome and not very chatty. However, the diffuse nature of the molecules and atoms in such clouds makes it unlikely that such a form could forge a self-replicating system that would evolve or be sustained over long time spans, let alone over the lifetime of a galaxy.

A compelling question—must life use water as a solvent?—has intrigued biologists for decades. In thinking about the answer, we consider contingency in the most basic requirements for living things—the liquid in which its parts are assembled. We continue our odyssey into the physical principles that shape the molecular level of life. Although water may look simple, a mere atom of oxygen bolted on to two atoms of hydrogen, this image belies its essential role in living things and the stunning variety of physics that explains life's attachment to the substance.

We know of no single organism that can be active without water, and we know of no form of life that can use an alternative solvent to do the bulk of its essential chemical reactions. The question is whether this requirement for water results from one very specific set of evolutionary conditions or whether it derives from something more fundamental.

It has long been recognized that water has some very unusual properties. One of the most notable to you and me is that when it freezes, ice floats on water, since the frozen water becomes less dense, a simple observation that you can verify by observing ice cubes in your chilled drink. This property is strange, but not unique; silicon displays a similar behavior at a pressure of about twenty gigapascals. Most liquids, however, when they become solid, become denser and sink in their corresponding liquids. Water has this apparently anomalous behavior because the individual molecules in the liquid link up with each other in hydrogen bonds. The oxygen atoms in one water molecule line up with hydrogen atoms in other molecules, a result of the polar, or bar-magnet-like, qualities of water. In the liquid state, water molecules are agile and move around freely. They can get close to each other and twist and turn to fit in the nooks and crannies. However, when frozen, those hydrogen bonds rigidify, forming a well-ordered network that, because of its regular structure, takes up more space than the liquid does. With its more spaced-out structure, ice becomes less dense than liquid water and it floats.

Because of this unusual behavior, ice remains on the surface of a frozen winter pond, leaving the fish underneath protected inside their watery habitat while everything above them is solid. The icy roof traps the heat beneath, slowing the further freezing of the pond and allowing the fish to enjoy a bird-free existence, at least until spring. These village pond observations have led many to gasp in amazement that the physics of water seems so well tuned to life, for if ice sank, the village pond would freeze from the bottom up, killing the fish. However, we should not jump to conclusions on the vital nature of this uncanny and apparently life-supporting property of water.

In the forests of North America lives an enchanting animal, the wood frog, *Lithobates sylvaticus*. It inhabits the undergrowth and, to the casual observer, looks like nothing special. However, come winter, the little creature has a fiendish trick. When the winter frost arrives, the frog buries itself underground in leaf litter and soil and, in a feat of biochemical wizardry, produces the sugar glucose in its bloodstream. The sugar prevents the blood from freezing, curtailing the formation of ice crystals, whose long, slender shapes might otherwise rupture the blood vessels and damage the frog. When spring returns, the frog warms up and hops off into the undergrowth, unfazed by the turn of events.

The ingenious wood frog is an example of how we should be cautious in how we view the world. The fish in the frozen village pond might well suggest that the unusual properties of water are just right for life, but the wood frog shows us that if life evolved in a fluid that froze solid during the winter, life could probably adapt to these conditions. The properties of water are not adapted to life; life adapts to its surrounding chemical and physical conditions, including the fluid in which life happens to exist. However, this observation still does not answer the question of whether water has properties that make it a unique solvent to provide the crucible for living things to emerge.

Some aspects of water make it far from ideal for life. If we dig hard enough, we can even find properties that are deleterious. The substance looks innocuous enough in your glass, but water is not inert and it has an unpleasant ability to react with some of the key molecules of life. Hydrolysis reactions, as you may have gathered from the root of this word (from the Latin *hydro-*, or *water*), are reactions in which water can cause chemical changes.

In the liquid state, water is not merely the familiar formula H_2O, but breaks up to form hydroxide ions, OH^-, and hydronium ions, H_3O^+, which are protons (H^+) bound to water:

$$2H_2O \leftrightarrow H_3O^+ + OH^-$$

The ions formed from the dissociation of water in this way can attack the long chains of life. From nucleic acids to sugars, hydrolysis reactions can cause these essential molecules to break apart, forcing life to use energy to constantly repair and rebuild itself against this damage.

Water may not be perfection, and with this contrarian outlook, we can certainly find aspects that speak against it. However, these minor quibbles aside, it has remarkable properties used in life.

Because its constituent atoms are slightly charged, or polar, liquid water can dissolve a wide range of small and large molecules—important for dissolving all those substances involved in the complex cascade of metabolic processes of life, from ions to amino acids.

Proteins—the diverse set of molecules that include the biological catalysts, enzymes, and many other pieces of biochemical machinery—have an uncanny and astonishing diversity of uses. Here, we see the true character of water and can view, in its full glory, what makes water such a good playground for the chemistry of life.

By binding to the outside of proteins, water molecules help keep them flexible, enabling them to move around sufficiently to take up the ingredients of the chemical reactions they drive as biological catalysts, but ensuring that they have enough rigidity to fold properly, maintaining their integrity. Strangely, in this role, water, often thought of as essential to maintaining stability, actually assists in destabilizing the protein just enough to encourage fluidity, showing the fine balance it oversees in life.

In other proteins, water molecules shield amino acids, preventing them from binding too strongly to other amino acids. This behavior, although apparently preventing stabilizing bonds from forming, again encourages just enough instability for the protein to remain flexible.

Even stranger collusions between water and proteins have been reported. The water molecules attached to the surface of a protein, because of their hydrogen-bonded network, become a "shell" of tightly bound molecules encapsulating the molecule. It is a physical state a little like glass. This behavior too plays a vital role in holding proteins together, but also in ensuring the easy movement of many of them.

In all these astonishing ways, water helps proteins to fold and enables their floppy chains of amino acids to coalesce in the right way. But it even goes further than this. Water can become part of the very structure of proteins, defining the shape and function of the overall molecule. By binding to the interior of the so-called active site, the region where the chemical reactions occur, water can link up with the incoming molecules, facilitating the catalytic roles that proteins perform. Water molecules are very much part of the way that many proteins work.

Not content with just getting into proteins, water deftly instantiates itself into the very code of life. How water molecules bind to DNA depends on the sequence of nucleotide letters within the DNA itself, so if the water molecules bound to the DNA then meet up with other parts of the DNA or other molecules in the cell, they are thought to mediate biochemical alterations related to the DNA code beneath them. This arrangement allows the genetic code to be read in an entirely unconventional way through the medium of water.

Water's roles in cell biology go beyond helping structures form and orchestrating important reactions; cells also take advantage of the liquid's ability to move electrons and protons around. Long chains of water, hydrogen bonded and behaving like wires, conduct protons around in bacteriorhodopsin, the molecule responsible for photosynthesis in some bacteria, allowing them to gather energy. In this clever arrangement, we see how the movement of subatomic particles through water is vital for some living things to acquire energy.

It is easy to dismiss some of this evidence as fascinating but a distraction from other possibilities. Much is made of the fact that some proteins can operate in non-water liquids such as the organic solvent benzene; this capability of proteins suggests that biochemistry could evolve in other fluids. However, most of these proteins must first be folded in water before they can do tricks in non-water solvents. The ability of some proteins to operate in organic solvents does not demonstrate that a whole biochemistry, with its many molecular interactions, could occur in other liquids or that even if it could, it is evolutionarily likely to happen elsewhere. Even when proteins are operating in nonaqueous alternatives, water is often still bound to these proteins and involved in their structural arrangement.

These different uses of water, in their intricate and manifold varieties, have driven home the point that we cannot simply think of life wallowing around in its solvent, but that the fluid is a fundamental part of the biochemistry of life. Life and its liquid are interwoven with such complexity and subtlety in so many ways that water is part of the machinery, not merely a medium in which other life-giving reactions happen to occur.

The impressive versatility of water, its diverse personalities that span across electron shuttle to proton wire, from hydrogen-bonded network to purveyor of molecular rigidity and flexibility, suggests that it may be unique in its capacity to integrate into, and play a major role in, a self-replicating, evolving living system.

Despite the growing list of water's impressive attributes, what we know about other liquids should give us pause for thought. One most popular solvent thought to be a possible alternative for water in life is ammonia (NH_3). At the equivalent of one Earth atmospheric pressure, its liquid temperature range is −78 to −33°C, but if you pressurize it, the boiling point can be raised to about 100°C, like the wide temperature range for water. Ammonia, like water, can dissolve many small molecules and ionic compounds too. The liquid might offer the potential of an environment for life where cold liquid ammonia solutions are thought to exist, such as in the deep subsurface of Saturn's moon Titan, in the atmospheres of gas giants such as Jupiter, or maybe in the oceans of icy moons. However, there the similarities with water end.

An essential characteristic of life is the ability to compartmentalize molecules from the outside environment using membranes. Liquid ammonia does not support the spontaneous formation of membranes in the same way that water can, although at low temperatures, hydrocarbons, including some lipids, can be separated in ammonia.

Part of the way ammonia's behavior differs from water's is that ammonia cannot form such strong hydrogen-bonded networks. This difference explains ammonia's lower boiling point—the molecules are more easily pulled apart when they are heated. Many of water's subtle interactions, including that fine balance of stabilizing and flexibility-causing effects in proteins, may not be as readily possible in ammonia.

To top it off, ammonia can aggressively attack molecules in life. Ammonia, like water, dissociates into two ions in solution (NH_4^+ and NH_2^-). This solution containing NH_2^- binds with protons and so attacks molecules that contain them. These molecules include a vast number of complex molecules from which the life we know is assembled. This annihilative behavior makes ammonia damaging to life on Earth and in all probability very reactive to many complex molecules elsewhere. To summarize in colloquial terms, ammonia just lacks chemical subtlety.

It would be remiss if I did not point out that ammonia has some strange and noteworthy properties, though. For example, it can dissolve metals, making an eerie blue solution of metal ions and many free electrons. Free electrons are essential in life because they are the raw material for the electron transport chains that gather energy from the environment. Superficially, we might claim that a liquid that can dissolve electrons would offer a source of this most sought-after commodity. Eerie blue aliens absorbing tasty electrons from their environment in oceans of ammonia? Let us not discount it.

Despite what I have discussed so far, ammonia can be involved in complex chemistry. It is used as a solvent by industrial chemists to prepare many useful things needed by industry. It is a precursor to a smorgasbord of nitrogen-containing compounds such as hydrazine, used in rocket fuel.

The problem with ammonia, like all nonaqueous solvents proposed for life, is that we can come up with a shopping list of characteristics that are favorable for a living thing. Most liquids have some properties that do not seem inimical to life and can be, as with the solvated electrons in ammonia, possibly even useful to it. However, we are not looking for a solvent with a few things that are compatible with self-replicating, evolving organisms. We are looking for a fluid that can participate in a vast diversity of reactions and that, with the sheer breadth of chemistry we would like for building a living thing, is not too blunt-edged or reactive in its chemical behavior.

We leave the oceans of ammonia to move to other liquids that seem even less likely as useful solvents but that, in our quest to be open-minded, we might consider anyway. Some have something positive to offer scientists' tireless quest for other liquids. Among these liquids are sulfuric acid (H_2SO_4), formamide (CH_3NO), and hydrogen fluoride (HF).

In contrast to water, liquid sulfuric acid has a much wider temperature range, from 10 to 337°C at one atmosphere pressure, which might make it look promising since it could exist in a liquid state in a wide collection of environments. It can be found in the clouds of Venus at concentrations between 81 and 98 percent. Intriguingly, within the Venusian clouds, at around fifty kilometers high, there is a region in which temperatures are within the range of 0 to 150°C and pressures are similar to those at the surface of the Earth. The optimistic temperature and pressure data alone have invited much discussion about the possibility of life, in the shape of floating bladders bobbing along in the Venusian sky or sulfate-reducing bacteria chomping on sulfuric acid. In an intriguing thought experiment, chemist Steve Benner suggested some alternative chemistries for proteins in strange liquids. In sulfuric acid, a link between amino acids could be made stable with a sulfur atom, instead of nitrogen. Although, like water, sulfuric acid can dissolve many compounds, it is not kind to organic material or to much complex chemistry. Its chemically destructive nature means that any biochemistry that evolved in it would likely be very limited.

Similar chemically limited vistas are found in formamide. Although many molecules, including some familiar to us, such as ATP, are stable in the substance, the smallest amount of water in combination with formamide hydrolyzes them, destroying them, meaning that oceans of formamide would have to be on an almost-waterless planet.

Not dissimilar in chemical character to water, hydrogen fluoride can form hydrogen bonds and will dissolve many small molecules. However, it is alarmingly reactive when mixed with water to form hydrofluoric acid. In the laboratory, geologists use this acid to dissolve away rocks, etching out fossils to make them easier to see. It's propensity to react with carbon-hydrogen bonds and turn them into carbon-fluorine bonds might stymie some of its attractiveness as a solvent for organic chemistry, unless a life form can be built with a fluorine-rich set of molecular ingredients.

In addition to the challenges just discussed, other problems may come into play with the theoretical alternatives to water in our search for life-supporting fluids. This is particularly the case for those liquids that are proposed to operate at low temperatures, such as liquid ammonia.

Chemical reaction rates proceed according to a very simple principle expressed in the Arrhenius equation. Svante Arrhenius, a Swedish Nobel Prize–winning chemist and physicist, was an extraordinary polymath of the nineteenth and early twentieth centuries. He dabbled in a vast number of subjects and even speculated on the effects of adding more CO_2 to the Earth's atmosphere, predicting that it would prevent ice ages and warm the Earth. He recognized that the rate of chemical reactions depends on their temperature. Using the rates of different reactions measured in the laboratory, he showed that this dependence was not a simple linear relationship. Doubling temperature does not just increase reaction rates by the same amount regardless of temperature. The relationship is exponential. More precisely, the rate of any reaction (k) is given by

$$k = Ae^{(-E_a/RT)}$$

where e is a mathematical constant, E_a is the activation energy, R is the universal gas constant, and T is the temperature
of interest. The more unusual factor, A, is a constant for each
chemical reaction; it defines the frequency of collisions having the correct orientation.

What does this exponential relationship between temperature
and the reaction rate mean for life?

Consider a reaction with an activation energy of 50,000 joules,
which is the energy needed to get the reaction going. Drop the temperature of the environment from 100°C to 0°C, and the reaction rate
decreases by just over 350 times. However, drop the temperature by
another hundred degrees, from 0°C to –100°C, and the reaction rate
decreases by a staggering 350,000 times! At the temperature of liquid
nitrogen (about –195°C), the rates of reaction are 10^{23} (100,000 billion billion) times less!

The optimistic might immediately hit back with the rejoinder that
catalysts could accelerate reaction rates, but even the best enzymes
and chemical catalysts increase reaction rates by only a few orders
of magnitude. This exponential relationship may not be a problem:
life can merely operate at these slower rates, maybe replicating many
times less frequently than typical Earth life. However, in most planetary environments, life is subjected to constant damage that must be
repaired. One source of this damage is background radiation.

Thus, life is confronted by a problem. It must be able to repair
radiation damage to prevent the damage from accumulating to fatal
levels. In the deep subsurface of Earth, where there is little energy
available to grow and reproduce, microbes might divide very infrequently. And yet even here, they must get enough energy to repair
damage from radiation. In the Earth's rocks, natural background

radiation would kill the most radiation-resistant microbes known after about forty million years if they remained dormant. On Mars, as the atmosphere is thinner than Earth's, the surface has the additional problem of higher levels of cosmic radiation. Here, even a radiation-resistant dormant microbe, if such a thing ever existed there or is accidently dropped there by human or robotic explorers, would be killed within thousands of years, or much less.

If the chemical reaction rates in the cold-temperature life form are many thousands, millions, or trillions times less than those in the life we are familiar with, it is likely that this cold life form will accumulate a great deal of damage and be unable to repair itself sufficiently fast to remain alive.

There may, however, be some more optimistic news for the low-temperature life form. Some challenges it faces depend on temperature. The formation of reactive oxygen species, the decay of amino acids, and the thermally caused decay of DNA base pairs depend on temperature: the lower the temperature, the slower the damage occurs. Although the low-temperature life form may well incur damage, some of this degeneration will be correspondingly slow, partly making up for the slow rate at which it can repair itself. However, direct damage to molecules caused by radiation can be essentially independent of temperature. Live life too slowly, and you may not keep up with this inevitable damage.

Besides slow repair and growth rates, the environment may run away from our sluggard creature. Any environment alters over time. Indeed, for chemical reactions to yield energy for life, there must be turnover and dynamic alteration in the environment to provide these contrasts. At extremely low temperatures, cellular chemical reaction rates are such that by the time a cell has commanded its metabolic pathways to take advantage of some short-term change in its habitat, there is a good chance that the conditions will have shifted again. On the larger scale, there is the problem that if reaction rates are reduced by many trillionfold, then it is likely that conditions on a planetary scale will have transformed before metabolic pathways have had a chance to respond to the initial conditions to which they were exposed. Life would be engaged in a futile catch-up game, trying to capture energy sources or respond to physical and chemical conditions long since gone.

There is likely to be a temperature range optimal for living things. Life's capacity to adapt and repair itself probably has a temporal correlation with the rate of radiation and other geochemical and geological perturbations in most environments in the universe. At extremely low temperatures, the processes of life may be generally out of sync with many processes that happen in and on planets.

It always helps, when thinking about the prospects for alien chemistries, to find real places in the Universe to test ideas. Many of the best-known examples of frigid environments in our own Solar System, such as the oceans in the icy moons of the gas giants or the glaciers of Mars, may not be much colder than places we know on the Earth. However, even in our own cosmic backyard, we do know of locations where there are liquids at temperatures significantly lower than anything found on Earth. Do we find any reason to be optimistic that they might harbor self-replicating evolving systems of matter?

There is a cold place in the universe where people have entertained the idea of life—Saturn's moon Titan. The surface of this remarkable moon was presented to a stunned human audience by the Cassini spacecraft and its lander, Huygens, which in 2004 returned jaw-dropping images of this ethereal world as it descended through the atmosphere. Rivers of methane carve through a landscape, creating sinuous tributaries and lakes much like the features of our watery world. In this frigid landscape, at a chilly −180°C, water ice behaves physically like rocks on our planet.

On Titan, the solvent on offer to life is methane, whose behavior as an organic molecule differs greatly from water. Unlike water, methane has no polarity, and so it has difficulty dissolving many ions and charged molecules that are vital in terrestrial biochemistry. Most proteins we know would be ineffective.

Some say that one advantage of methane is that it is less reactive than water. Those hydrolysis reactions that damage life's molecules on Earth would be nonexistent. Although this may be the case, water's tendency to react with some molecules is an essential part of its ability to maintain molecular flexibility and to choreograph dances and communication between molecules. The reactive nature of water, although sometimes unpropitious for life, is usually beneficial for any living thing.

A popular line of argument is that some chemists actually prefer to do some syntheses in non-water solvents, where the reactive nature of water can be avoided, proof that life would be better off without water entirely and that liquid methane and similar fluids may be a step up for life. However, chemists like doing reactions in organic solvents since their objective is to maximize the yield of compounds they are trying to make. They want to lose as little as possible in unwanted, reactive chemistry. But this is not life's game. Life employs reactivity to drive active biochemical processes. Methane's comparative lack of reactivity and its inability to dissolve polar molecules is unlikely to constitute some sort of advantage to life in the same way that these properties might entice industrial chemists.

With some imagination, however, we can think of how to build biochemical structures in the presence of this organic compound. Consider how to make a membrane like those used by life on Earth to enclose a cell. One way we might make such a membrane on a Titan-like world is to flip it around, to make an inside-out membrane. The charged heads would point inward toward each other to escape the water-hating methane, into which the long fatty acid tails would point. By turning the lipids around, we could make a vesicle appropriate for a methane world. To make this work, we could not incorporate the fatty acid tails that life uses on Earth, as they would be almost solid and immobile in the cold methane lakes of Titan. Instead, using chemical modeling, a team at Cornell University invented a membrane formed from acrylonitrile, a nitrogen-containing compound known to be present on Titan. Their *azotosome*, as they call it, would have polar heads rich in nitrogen; the heads would be attracted to one another to form the membrane, with tails of short-chained carbon compounds sticking out. Using this chemical compound, the whole structure would maintain fluidity on Titan similar to membranes on Earth.

Not content with models and speculation, we might look at real data. Researchers have approached the possibility that Titan could host life by comparing measurements of gases on the moon with the ways that life might make energy. They proposed that by reacting hydrocarbons such as acetylene and ethane with hydrogen, present in Titan's atmosphere, life could make energy, producing methane as the waste product. These ideas have even received some boost from the observed depletion of hydrogen in the atmosphere of the moon and an apparent depletion of acetylene near the surface, suggested as tantalizing circumstantial evidence for life. These data are highly provocative. In applying Occam's razor, a principle that cautions us to accept scientific explanations with the minimum number of assumptions—a principle particularly important for thinking about alien life—we should remember that our still very limited knowledge of Titan and its methane cycle might well hide other geological and geochemical explanations for these observations. Nevertheless, the ideas are fascinating.

As we can see, with a little imagination, we can construct an internally self-consistent picture of life on Titan. However, the presence of possible energy sources for life, an abundance of organic molecules and other noncarbon atoms, and potentially even lipid-like compounds on Titan may not be sufficient for life if low temperatures in most of its lakes and landmasses prevent a viable living system.

As with all these discussions, we probably have some chemical bias because we focus our research efforts on the solvent we know so well—water. Our knowledge of ammonia, liquid nitrogen, hydrogen fluoride, liquid methane, and other solvents is less complete and as we have no example of a biochemistry that operates in them, we are left to a fair amount of speculation. If we ourselves used another solvent, could we predict how the strange solvent dihydrogen monoxide (H_2O, water!) could interact with a reproducing, evolving, self-replicating organism? Even though we are a water-based intelligence, our knowledge of its role in biochemistry has only recently advanced rapidly and yet remains in its infancy.

Nevertheless, even with this caveat, the solvent seems an incredibly versatile substance. Water has an extraordinary capacity to play leading roles and bit parts in the theater of life. As yet, other solvents hosting organic chemistry or even other biochemical architectures of life have not been shown to possess this multitasking capacity. Equally important, water is liquid in a range in which chemical reaction rates are commensurate with the need to deal with biologically damaging agents, such as radiation and changing conditions at microscopic scales right up to momentous rearrangements at the planetary scale. Alongside its chemical promise as a solvent for life, water has a cosmic copiousness, suggesting not only that its physics is suitable for living things, but also that the physics of the wider universe makes it a common solvent on offer for any emergent planetary experiment in evolution.

Twelve billion light-years away, there is a quasar, an ancient object with the rather unmemorable name of APM 08279+5255. Astronomers have a penchant for names like this. I'm a biologist. Let's call it Fred. Now Fred harbors a black hole about twenty billion times more massive than the Sun. Astronomers do not yet understand quasars. You will appreciate that since Fred is twelve billion light-years away, we are observing light from near the beginning of the universe. Quasars are very old objects. Nevertheless, within this curious distant fuzz, there is a vast quantity of water—a voluminous 140 trillion times all the Earth's oceans combined!

Fred is not unusual. Water can be found everywhere: it is a common volatile. In our own Solar System, we have the ocean under the ice cover of Jupiter's moon Europa and the jets of water—geysers—erupting from the south pole of one of Saturn's moons, Enceladus, an unpretentious moon less than five hundred kilometers in diameter. Then there are the ice caps of Mars and those frozen comets, about one to ten billion of these objects with a diameter greater than one kilometer in the Kuiper Belt alone.

How the water found in Fred was produced is a matter of some conjecture, but regardless, it is thrilling that astrochemists have schemes for how it might be formed in these very alien environments. Look at the reaction scheme below:

$$H_2 + \text{cosmic irradiation} \to H_2^+ + e^-$$

$$H_2^+ + H_2 \to H_3^+ + H$$

$$H_3^+ + O \to OH^+ + H_2$$

$$OH_n^+ + H_2 \to OH_{n+1}^+ + H$$

$$OH_3^+ + e^- \to \underline{H_2O} + H; OH + 2H$$

The chemical details need not concern us, but the simplicity is beautiful and worth remarking on. Molecular hydrogen is bombarded with some radiation, perhaps from a dying star. Some hydrogen ions are produced and can react with oxygen atoms, the oxygen itself having been produced and strewn across interstellar space in supernova explosions. Then the ions containing hydrogen and oxygen react with more hydrogen ions to produce an OH_3^+ ion that can mop up an electron and become water. I have underlined the water above.

So we take hydrogen from the big bang and some oxygen from exploding stars, mix in some radiation and electrons, and we produce water, everywhere across the universe.

These reactions may not be the only ones that give Fred its water, but they show that the pathway to water is simple and requires no special conditions. The water that Earth obtained in its early history, once thought to be from comets, probably mainly came from water-rich asteroids, the water within them having been originally formed in reactions like the ones described above. Fred tells us that these processes have been going on for billions of years. In one location in space, over seven billion years before Earth was formed, before life on our planet would emerge, trillions of oceans of water were produced around just one object.

The other solvents that have attracted attention as plausible candidates for life tend to be rarer. The ocean of water thought to exist under the surface of Titan may contain 30 percent ammonia, a water alternative likely to have been one of the components in early Earth's atmosphere. Today, ammonia is one ingredient of the atmosphere of Jupiter. The compound is out there, but probably not as abundant as water. Sulfuric acid, the more eccentric suggestion of a liquid for life, is even rarer. As for hydrogen fluoride oceans, they seem unlikely. Fluorine is about a hundred thousand times less cosmically abundant than oxygen. Whatever their chemical versatility, these alternative solvents and others fail to match the sheer quantity of water in the universe. Other fantastical life-giving liquids in the universe, oceans of sulfuric acid or ammonia in which fishy life forms swim, are likely to be much rarer than our comely water oceans. The physical properties of water make it both abundant and versatile as a solvent in which to assemble life.

CHAPTER 10

THE ATOMS OF LIFE

To BEGIN A CHAPTER of a book on life with an excursion into *Star Trek* may not seem like a very auspicious development, but this television and film series that began in 1966 from a concept by Gene Roddenberry is an example of that pervasive view that biology is limitless. Across the galaxy, the crew of the starship *Enterprise* gallivant around, encountering strange life forms and trying to figure out ways of defusing their often irascible tempers or aggressive tendencies. The theme of this series, that the universe contains a never-ending supply of unpredictable biological potential, is a common idea in science fiction. *Star Trek* constructed decades of television and film from this simple notion.

I absolutely deny being a Trekkie, but I agree with William Shatner, Captain Kirk's real-life incarnation, that the best episode made was titled "The Devil in the Dark," which aired in 1967. Fifty miners lie dead on a planet called Janus VI, apparently killed by an annoyed creature that sprays things it encounters with a corrosive substance. The creature is tracked down and turns out to be a silicon-based life form made of the same silicate substances from which rocks are assembled. The silicon nodules that the miners have been collecting (one nodule sits on the desk of the mine's director) are not mere boulders, but the eggs of these creatures, the Horta. After some reconciliation with the crew of the *Enterprise*, some Horta etch NO KILL!

into the rock, and with this sentient coming-together of cultures, the Horta help the miners locate precious metals in exchange for being left alone to tend to their eggs. Everyone couldn't be happier.

The Horta and their offspring reflect another fundamental question in biology: can the elements from which life is constructed, the atomic building blocks of life, differ from those we know on Earth? With this most basic of questions, we continue down into the hierarchy of life, now moving into the atomic scale to peruse how physical processes might shape and channel its construction at this more fundamental level of matter.

Life on Earth uses a vast range of elements as the atomic chassis of its basic molecules, but the predominant element that forms the backbone, if you will, of the vast pantheon of molecules that come together to building living organisms is carbon. This element occupies group 14 of the periodic table. Silicon, the element below it, belongs to the same group and shares similar chemical characteristics. So then, imaginative people ask, why couldn't silicon replace carbon in life as the next best alternative? With the universe full of silicon, there is no shortage of the substance for building living things. As Kirk might well have pondered, what's not to like about the Horta?

To answer this question and to explain why life has made the choices it has with all the elements it uses, we must know something about the structure of atoms from which life is assembled. By delving into the periodic table and the physics of atoms, we will find that extraordinary universal principles of physics ultimately lie at the heart of carbon as the favored element for constructing life forms.

Developed in one of its first modern forms by Dmitri Mendeleev in 1869, the periodic table contains all the known elements, either known naturally or synthesized in the laboratory. At the core of every atom of every element is a nucleus, which contains protons, positively charged particles. Besides hydrogen (with only one proton), the nucleus of all other atoms has some neutrons too. These neutral particles play a part in binding the nucleus together. Elements are enumerated according to the number of protons they have, sometimes called the *atomic number*. So hydrogen, with one proton, is element number 1. It sits at the top left of the periodic table, and the somewhat awkwardly pronounced oganesson, element number 118, sits at the bottom right.

Surrounding this small bundle of particles in the center of the atom are electrons, subatomic particles that also have something of a wavelike quality to them, like light. But unlike the protons, the electrons have a negative charge. Atoms are always neutral—they have no charge—so the positive charges of the protons must cancel out the negative charges of the electrons. In other words, the number of electrons in an atom must be the same as the number of protons.

So far, we have a simple picture of elements increasing in atomic number from the top left to the bottom right of the periodic table. To create each element, one proton in the nucleus and one electron metaphorically in orbit around it are added sequentially, building the zoo of atoms from which the universe, and life, is made.

There is a little problem in the view I have presented. We cannot just add electrons to the atom one by one in a growing crowd of the particles. Electrons hate being next to other electrons which are exactly the same, a little like identical twins dumped next to each other at a birthday party—twins who dislike being compared and who prefer their friends to treat them as distinct. Therefore, you can't just stack electrons next to each other. This principle, that electrons, or all fermions, cannot occupy the same states, is called the *Pauli exclusion principle*, named for the inventor of this concept, Wolfgang Pauli, an Austrian-born physicist.

What then, do we do with two electrons that are side by side in an atom and do not want to be identical? One property we can alter is their spin. If the electrons' spin is in different directions (spin up and spin down), then they are now distinctive. Like the twins that have some distinguishing feature from which they can feel a semblance of individualism, the two electrons can now abide by the Pauli principle. However, this principle prevents us from adding a third electron, since there is no other property we can modify to make the third electron different. Like a metaphorical subatomic Noah's Ark, the electrons are stacked into the atom two by two.

As we add electrons to atoms, they occupy so-called orbitals, sometimes called *shells*. Each shell or orbital can contain two electrons or multiples of two electrons, ensuring that the Pauli exclusion principle is not violated.

Once the stacking is completed, those outermost electrons that filed into the last orbitals are of singular importance because they are the part of the atom that will first come into contact with another atom; they will define the nature of any chemical bonds or whether atoms will react with one another at all. Atoms that have partly full electron orbitals like to gain or lose electrons to end up with a complete set of electron pairs; empty electron positions make atoms reactive.

Pauli's little rule explains why the noble gases, such as neon and argon, are famously inert. They have their outermost electron shells full to the brim with electron pairs—four pairs of two electrons, with no room to spare—meaning there are no spaces left to accept electrons from other atoms or to participate in exciting chemical reactions. The noble gases are left conservatively unreactive.

This stacking of electrons in atoms explains how the elements, from 1 to 118, are arranged in the periodic table. Each column of the table contains the atoms that have the same number of electrons in their outermost shell. This means that within each column, because they have the same number of outer electrons, the elements have very similar chemical properties. So now you can see that the properties of atoms and the way they shape the material world around us are decided by the way the electrons are all stacked. That is determined by a simple physical principle: the Pauli exclusion principle.

Let us return to life and consider the element at the core of most of its molecules: carbon. It has six electrons. Those six electrons must be stacked in a way to keep Pauli satisfied. Two electrons sit in the so-called 1s orbital, the lowest orbital. Two electrons can then be stacked into the next orbital up (the 2s orbital). The remaining two are in another orbital at the same level, the 2p orbital.

What about the Horta? The speculative creatures are made of silicon, which is in the same group of the periodic table as carbon is, but one row down. Silicon contains fourteen electrons. How are they stacked? Two electrons are in the 1s orbital, and two in the 2s orbital, like carbon. Six are then stacked in the three 2p suborbitals. Two are placed in the next orbital up, the 3s orbital, and then finally two in the 3p orbital. Although silicon has more electrons than carbon does, the two elements' outermost shells are very similar—two electrons in an s orbital and two in a p. This similarity explains why carbon and silicon have similar chemical properties and why the Horta came to exist in our minds.

Now that we have a grasp of one fundamental principle that lies at the heart of biology—at its lowest level of its hierarchy, the atomic level and its constituent subatomic components—let us explore a little more what makes carbon a good building block for life and whether silicon might suffice, too.

Carbon is just the right size. In its outermost shells, the electrons that stand ready to pair up with electrons in other atoms and form bonds, thereby forming molecules, are bound close enough to the nucleus that they hold on tight, meaning the links are strong. They are not so far away that they are pulled off the atom easily, which would make the bonds liable to break. Life must be able to build molecules that, like DNA, will be stable, but it must also be able to pull apart old molecules to make new ones without using vast amounts of energy. Carbon fits the bill.

The electrons in the outermost orbitals, those electrons in 2p and the other pair in 2s, love to pair up with electrons from other atoms and form bonds. In an especially common reaction for a carbon atom, one of its electrons connects with the single electron in hydrogen to form a carbon-hydrogen bond, a union that decorates all manner of life's molecules. Carbon can form bonds with other carbon atoms, and with sulfur, phosphorus, oxygen, and nitrogen too. These bonds have similar strengths, so carbon needs little energy to switch between these different atoms. The atom has other arrangements. It can form double bonds. The two electrons in the 2p orbital can pair up with two electrons in the 2p orbital of another carbon atom and form the double linkage. This capacity, along with an ability to form triple bonds, adds to the welter of carbon-containing molecules.

Resulting from all this versatility and enthusiasm to form bonds is an impressive diversity of chains, rings, and other structures: from the simple gas methane, made of just one carbon atom bound to four hydrogen atoms, to the amazingly long molecule DNA, which unraveled is a full two meters long in a human! It is therefore natural for anyone to ask, when confronted by this flexibility in the assembly of a range of molecules, if other elements could do the same. Silicon is an obvious contender, and as the second-most abundant element on Earth after oxygen, it might well look like a pretty good candidate.

Despite the similar electron configuration on the surface, there is one crucial difference between silicon and carbon. As noted, silicon has fourteen electrons stacked up, in contrast to carbon's six, which means that silicon's outer electrons are further away from the nucleus and less tightly bound than carbon's outer electrons. A consequence of silicon's more lightly bound electrons is that its bonds with other molecules tend to be weaker than carbon's. The silicon-silicon bond is about half as strong as the carbon-carbon bond, meaning that rarely in nature can you find more than three silicon atoms bound side by side. As a result, there is little chance that silicon can build all those complex chains and rings that we find in carbon-based life, where many dozens of carbon atoms can be joined in chains. The electrons, being less tightly bound to the nucleus, are more apt to be snatched up by other atoms or themselves to pair up with other electrons, making the atom more reactive. Some bonds that silicon forms are very unstable. Silane (SiH_4), an analogous molecule to the biologically important gas methane (CH_4), spontaneously combusts at room temperature.

However, silicon has another Achilles' heel. A carbon atom, when it binds to an oxygen atom, can form a double bond. With two oxygen atoms, the result is the very versatile gas carbon dioxide, which is the raw material of photosynthesis. However, silicon, because of its larger size, cannot so easily form a double bond with oxygen and instead forms four single bonds that are more comfortably distributed around the larger atom. These oxygen atoms still have a single bond to spare, and they use this to bind to another silicon atom. The result? A giant network of silicon and oxygen bonds linked to form a great grid. And this grid is very familiar to you and me. It is the structure of the silicates, the material that makes glasses, minerals, and rocks. Unfortunately, unlike many other silicon compounds, they are so stable that once silicon is locked up in these structures, it obstinately stays there. Rocks are one of the most visible reasons why silicon-based life is unlikely.

These rocky silicates can be found in a dazzling variety. In a crude way, they recapitulate the enormous treasure chest of carbon compounds. But silicates are the stuff of rocks, not biochemistry. Their networks make them inert, so unreactive that silicate ceramics are used as heat shields to protect spacecraft as they enter the Earth's atmosphere; the intense temperatures, rising well above a thousand degrees Celsius, are still unable to cajole the structure of the material to do something interesting.

Although most of the silicon on our planet may be locked up in generally unreactive silicates, life is by no means devoid of the element. Diatoms, algae that inhabit the oceans and freshwater rivers, lakes, and ponds, protect themselves inside a frustule, an ornate shell made of silica (silicon dioxide). These photosynthesizing microbes achieve a beautiful diversity in their forms, including stars, barrels, and boatlike shapes. Plants also gather up and use silica. In some, the amount of the substance may be up to a tenth of the plant's total mass! Silicon is readily absorbed from soil as silicic acid and is thought to play a part in growth, mechanical strength, and resistance to fungal diseases. It is prominent in phytoliths, silica structures that are formed in cells and that aid in the plant's rigidity, necessary for upward growth against gravity. Silica structures, known as spicules, are even found as a primitive skeleton in certain marine sponges, organisms that belong to some of the earliest multicellular organisms on Earth.

No sensible scientist would discount silicon as a basis for life. Even on Earth, where silicates form 90 percent of the crust, the element need not be entirely bound with oxygen in silicates. The silicon and carbon compound silicon carbide (SiC) occurs naturally. In the interstellar medium, many silicon compounds such as SiN (silicon mononitride), SiCN (silicon cyanide), and SiS (silicon monosulfide) have been observed, showing that on the universal scale, silicon can make some unusual compounds. We have a certain bias since we know so much more about carbon chemistry than we do about silicon chemistry. As we delve further into silicon chemistry, we come across some surprises. The atom seems to form a colorful variety of compounds in concert with carbon—the organosilicon compounds, some of which form chained arrangements. Perhaps a black-and-white view of the two elements misses the potential for some sort of hybrid carbon-silicon-based life form.

Give it a chance, and silicon can form more-promising products. Among its family of structures are the cage-like molecules with the tongue-twisting name of silsesquioxanes. All sorts of structures can be added to these molecules' core to produce a magnificent collection of other molecules. Other silicon compounds, under just the right laboratory conditions, can be enticed to form silicon chains with over twenty consecutive atoms, like the long-chained compounds that make up the molecules of living things.

Although these amazing jaunts through silicon chemistry show that there are complex silicon compounds that have astonishing multeity, life has not been idle. It has evolved to test out this element in many functions, but it has not yet, as far as we know, used silicon to build the major molecules of life to such a pervasive extent that we would describe an organism as silicon based. Silicon-filled plants still have cells constructed with sugars, proteins, and lipids, the stuff of carbon-based chemistry. Tellingly, when life gets hold of silicon, cells do rocklike things with it—building siliceous structural support materials such as phytoliths and spicules. While perhaps life's structural use of silicon is a vestige of Earth's evolutionary history of life, which chose carbon, if organisms found some benefit to using silicon in many compounds that enhanced their chances of survival, they would use it. Earth's evolutionary experiment shows that under the conditions associated with our planet, carbon trumps silicon in almost all biochemistry.

The other elements in group 14, the same group that carbon and silicon belong to, suffer problems as the atoms grow in size. Germanium is the next element down, but germanium life forms have never been entertained. As far as we know, this element is also unable to produce the range of chemical compounds useful for building living systems. Onward down into group 14, tin or lead Horta seem to have even less chemical evidence to support their existence.

For all the pushing and pulling we can do with the periodic table to find elements that might look like good choices for fashioning life, carbon remains far and away the element with the largest and most diverse number of molecules in its repertoire of bonding possibilities. It is likely that the processes of life elsewhere in the universe would converge on this element as the basic elemental building block of life. And as described earlier, carbon is the best choice because of the Pauli exclusion principle, a universal principle operating at the quantum level and laying down rules for how electrons are stacked within atoms.

Those people with a healthy skepticism might still be unconvinced. Could other life forms modify not only the core atoms that they use, but also the solvent in which life operates? Maybe our assumption that carbon chemistry is linked to water limits our capacity to imagine alternatives. Should we consider a different union between the liquid and the central elements of life? An imaginative, if highly unfamiliar suggestion is that silicon-based life could originate and evolve in liquid nitrogen. The liquid nitrogen would offer sufficiently cold temperatures for complex and otherwise generally unstable silicon compounds, such as silanes and silanol compounds, the latter analogous to alcohols on our own world, to remain stable.

A wild and extraordinary geological cycle can be concocted from these suggestions. Reactions of silica in the rocks of a planetary core with carbon dioxide, ammonia, and other compounds would produce silanes and silanols. These would eventually be transported into the liquid nitrogen ocean, where they could participate in further chemistry, providing the basis of a silicon-based life. A location for such a novel type of biology has been suggested to be Neptune's moon Triton, an ice-covered world with nitrogen geysers on its surface, possibly erupting from deep, buried, frigid liquid nitrogen just beneath its surface. However, any place with some rocks and liquid nitrogen will do for this bizarre system of life.

These types of exotic chemistry and solvent combinations are even more difficult to assess than a relatively simple swapping out of one atom for another, since they stray into the realms of chemistry we know little about. The full capacities of silicon chemistry in liquid nitrogen are quite unknown, and we cannot rule out life in such circumstances in light of our knowledge of chemistry alone.

There may be good reasons, though, for seeing carbon in a positive universal light, even with these interesting alternatives on our minds. Not only does carbon have promising atomic physics for building complex life forms, but this propensity to form multitudinous molecules ensures that carbon molecules are abundant in the universe, making it likely that other examples of evolution, if they exist, would from their earliest stages find carbon molecules the most readily available larder of complexity.

Look up into the sky on any clear night, and you see a universe that billions of pairs of eyes have gazed upon throughout the history of our civilization. A canvas of black interspersed by the glinting white points of celestial bodies. Every now and then, their unchanging positions are interrupted by a comet, the bright glow of a supernova, or the nightly streaks of debris burning through our atmosphere as shooting stars, but otherwise the night sky seems immutable in human life spans.

This view of space as an endless landscape of emptiness is not inaccurate, at least compared with the rich variety of matter that packs onto our small world. However, to believe that the vastness of the universe is barren, a view that has dominated our collective consciousness since we first realized those points of light were stars and the blackness between them the rest of the vacuum of the cosmos, would be to overlook the breathtaking complexity of chemistry that goes on in this apparent void.

In the beginning, when the big bang heralded the beginning of the universe, things were simpler. As temperatures dropped, chemistry was confined to a few reactions between hydrogen, helium, lithium, and their ions, with some electrons and radiation thrown in for good measure, a basic playground of elemental rearrangements. Then the first swirls of gas gravitationally collapsed to a sufficient density to trigger the fusion reactions of stars. Within these glowing balls, the joining of hydrogen atoms into heavier elements, carbon included, could occur.

Some of these stars, the so-called low-mass stars, fizzled out, eventually burning through their fuel and collapsing into the gentle years of stellar retirement as white dwarfs. But some of the stars, more massive and within which more-elaborate onion layers of elements reside, collapsed in a violent conflagration, the gravitational forces catastrophically overwhelming the pressures of gas and thermal energy pushing outward, causing the convulsive shedding of material. In these massive explosions, or supernovas, new and heavier elements beyond iron in the periodic table were forged and strewn across the universe.

From astronomers' early work in the nineteenth and twentieth centuries, there emerged a better appreciation of where the necessary elements for life came from. Many light elements, including carbon, were mainly fashioned within the cores of low- and high-mass stars, while many other heavy elements needed by some types of life—molybdenum and vanadium, for example—were synthesized within supernovas.

This understanding of how elements were formed was a startling advance in our perspective of how life fits within the cosmic context. With this astronomical insight, we could now know the very origins of the elements of life and thus link the physics of the universe to the atomic structure of living things. Within this growing clarity, there was also something strangely disturbing and sobering about the truly cosmic origins of our existence. It has become something of a hackneyed observation to say that we are all stardust, but perhaps that sense of triteness merely reflects how much we take for granted our modern grasp of cosmology and astronomy. The ancients would have found such a statement baffling and abstruse.

This appreciation of our understanding of the link between life and the vacuum beyond might have been the first phase of a truly astrobiological comprehension of our origins. It was within the second half of the twentieth century, running through to the modern day, that another phase opened up and allowed us to grasp the ubiquity not so much of the elements for life in general, but specifically of the universal nature of carbon chemistry.

We can turn our telescopes toward that black void, but instead of viewing it in the range of the spectrum you and I are familiar with (the visible region), we can view it in the region just beyond the red—the infrared—using sensors that can turn that data into something you and I can see. If we study the infrared data, we observe not blackness but the swirls and endless beautiful puffs and eddies of gas, giant clouds that billow with grand vastness across the night sky. Where there was blackness, we now observe material.

Much of this now-visible material is *diffuse interstellar clouds*. They are so named because in their interiors, the gas concentrations can reach down to about 10^8 molecules or ions per cubic meter. Now this might sound like quite a lot, but all around you right now, there are about 2.5×10^{25} molecules of gases per cubic meter that make up the air you breathe. The diffuse interstellar clouds contain less material than do the vacuums we can create in laboratories on Earth. Nevertheless, within these clouds, there is still enough material for some astounding chemistry.

Recalling chemistry lessons at school, you might remember that to make a chemical reaction happen, you must get a high concentration of the reactants. Add very dilute sulfuric acid onto household sugar, and nothing much happens. Excitement only mounted when you entered the classroom to see the bottle of viscous yellowish concentrated sulfuric acid on the chemistry teacher's desk. The evil grin of the teacher gave away the black volcano of classroom distraction about to be unleashed from the dish of sugar soaked in the acid, and the acrid fumes as the sugar molecules were violently disassembled fulfilled every health and safety officer's dream of form-filling. So, you might ask yourself, how could anything of any interest be going on in a space cloud that is more diffuse than a typical lab vacuum?

There is something that occurs in abundance in space, but is not easily found in the school classroom: radiation. Protons, electrons, gamma rays, UV radiation, and many heavy ions such as ions of iron or silicon permeate interstellar space, including our cloud. This radiation imparts energy to the meager collection of ions and molecules, but nonetheless at sufficient levels to break them apart, energize them, and drive reactions between different kinds to produce novel chemical compounds. Even though these clouds are at a chilly temperature of about −180°C, the radiation can bombard the ions and molecules to force the chemical reactions along their way.

Astronomers can observe these products of interstellar chemistry using spectroscopy. As light passes through a diffuse interstellar cloud, compounds will absorb some of that light. More exactly, electrons will absorb the energy and jump energy levels, essentially introducing a ghostly void, if you will, in the spectrum of the light by robbing it of particular wavelengths. By analyzing the spectrum that passes through the cloud and reaches telescopes on Earth or in space, scientists can use this absorption spectroscopy to identify what is in these clouds. Alternatively, if electrons absorb energy from light and reradiate it, they emit light, perhaps of a different wavelength, which again is a fingerprint of a particular compound. Both of these spectroscopic methods allow us to build up a family portrait of the chemical compounds that make up a cloud. Our understanding of the results of these sophisticated approaches, though, is still very much in its infancy. There are countless absorbances and emissions in the clouds, whose origins we understand only vaguely or not at all. Diffuse interstellar bands, an assortment of absorbances found in spectra from clouds, contain many signatures that we are still at a loss to explain.

Despite the mysteries that remain lurking in these clouds, what has been successfully identified is an enormous variety of simple compounds, including CO, OH, CH, CN, and CH^+ ions. Now you will notice that within this short list—and there are many more compounds besides—one thing does stand out: there is a great selection of carbon-containing compounds. Carbon, produced by fusion in low- and high-mass stars and eventually flung into space to later coalesce in the clouds, reacts with many other elements to form simple compounds that contain the nascent structures of organic carbon compounds.

Things begin to get decidedly more interesting if we turn our attention toward other objects in the universe, for bigger and mightier clouds pervade the realms of the cosmos. Here and there, we can observe *giant molecular clouds*. These objects are gigantic, some about 150 light-years across. They can contain a thousand to ten million times the mass of our own Sun. These are the nurseries in which new stars are formed, the density of gas sufficient for eddies and swirls to congregate and initiate fusion burning of an infant orb. The density of material is much higher than in the diffuse interstellar clouds, some one trillion ions or molecules in a cubic meter, still much less than the air you breathe, but now sufficient to make chemistry that is even more interesting.

The material these clouds contain is now dense enough to shield much of the UV radiation being given off by new stars and other astrophysical objects. Although there may be less radiation to drive chemistry, it means that the compounds formed are less liable to be broken up by that same radiation. Within giant molecular clouds, we find over one hundred chemical compounds, including $HCOOH$, C_3O, C_2H_5CN, CN, CH_3SH, C_3S, NH_2CN, and many more. The picture should be becoming clearer. Within these cosmic nurseries, we go from very simple molecules containing one or two atoms to more impressive structures. The startling observations in our diffuse interstellar clouds are magnified—the complexity of carbon chemistry has increased. Giant molecular clouds are full of carbon-based chemistry!

The complexity that clouds can harbor is astonishing, and beyond compounds with just a few atoms, there are more startling structures. The six-atom rings of carbon, including benzene, can be attached together to make a family of molecules called *polycyclic aromatic hydrocarbons*. Intriguingly, the six-atom carbon rings can be reacted in the laboratory to form quinones, molecules involved in shuttling electrons around in life as it gathers energy from its environment. These sorts of laboratory reactions provide tantalizing demonstrations that molecules within the interstellar medium are already some way along the path to being useful precursors in the energy-yielding and metabolic pathways of living things.

These joined carbon sheets are thought to make yet more-unusual molecules. Assemble carbon rings in three dimensions, and you can create carbon balls, such as buckyballs that contain sixty carbon atoms. These C_{60} soccer-ball-like compounds made of twenty hexagons and twelve pentagons of carbon atoms linked together into a sphere can themselves coalesce to form layered onion-like carbon structures. They form tubes and grids of interlinked carbon atoms, exploring many possible combinations of structures.

Despite the revolution that has come from these observations in our grasp of how chemistry operates and what products it forms at the universal scale, scientists were still flummoxed by how this diversity of chemical compounds might have formed. They were especially perplexed by two important issues. To begin with, chemical compounds must be close to each other to react. This observation is underscored by the aforementioned example of sulfuric acid. Mix it with water, and the molecules of sulfuric acid become so dilute they do little exciting chemistry. Mix this diluted acid with sugar, and we end up with slightly acidic dissolved sugar, but no exciting classroom drama. As the molecular clouds are so diffuse, compared even with air, how could any chemistry possibly occur? And things get worse. The clouds are cold, very cold. We also know from our chemistry lessons that heating is a good way to get a reaction going. A small strip of magnesium metal does very little lying around on a lab bench, but place it in the Bunsen burner flame, and as long as the temperature gets up to above 473°C, it will ignite in a bright white glow. At −260°C to −230°C in the molecular cloud, the possibility of getting chemicals—or chemists—excited seems decidedly thin.

Yet chemical reactions occur, and they do so in surprising places. Scattered through the clouds are particles, interstellar dust grains that contain a core of silica or carbon-rich material surrounded by ice. The ions and molecules within a cloud can attach to these grains and concentrate. Now we have a mechanism to bring them together; otherwise, these ions and molecules would be left floating aimlessly in interstellar space. Astrochemists think that much of the chemistry in molecular clouds happens on these grains. Each grain is a factory and, more interestingly, a miniature reactor for making organic compounds.

The profusion of carbon in the universe is so great that some stars are even defined by the element. At the edges of "carbon stars" is a plethora of reactions that form, destroy, or shuffle carbon, generating an unknown level of complexity and diversity in organic chemistry. Just 390 to 490 light-years away, there is a star that can be seen to have halos, a signature of the envelopes of material expanding away at over fifty kilometers a second. Within the envelopes, over sixty types of molecules have been detected, among them linear and cyclical carbon molecules, organic chemistry literally being blown into space. Around the photosphere of the star, the region that emits light, simple compounds such as CO (carbon monoxide) and HCN (hydrogen cyanide) have been detected.

These rather cursory observations are sufficient to make some points that are probably becoming evident. The universe is not a cold, chemically infecund place where stars listlessly revolve around the center of galaxies, merely producing and spreading the basic elements of the periodic table, which the origin of life and eventually living things must assemble through unfathomable and obscure mechanisms into the organic building blocks for a reproducing, evolving entity. Everywhere throughout the universe, complex chemistry is occurring, even in the most diffuse puffs of gas, and this chemistry includes a vast pantheon of organic chemistry, generating bonds between carbon and other elements in a huge panoply of arrangements.

We see within this emerging picture a strange inevitability in the production of complex carbon chemistry. The variety of possible ways carbon can bind with other elements and produce a great catalog of molecules is not an idiosyncratic product of the temperature and pressure conditions found on Earth, a limited case of a specific planetary environment that allows carbon to take center stage in the world of complexity to build living things. In the coldest of places in the universe, carbon still does its thing, binding with the elements of the periodic table, including itself, to produce the gelateria of flavors of organic chemistry from which life on our planet is constructed. The path to the building blocks of carbon-based chemistry seems universal.

No doubt, the greatest interest from our perspective is how far this chemistry can go to assembling the molecular precursors of life. Attempts to find the basic monomers from which life is constructed, such as amino acids, sugars, and the nucleobases that make up proteins, carbohydrates, and the genetic code, respectively, have yielded mixed results. Researchers from Copenhagen University found glycolaldehyde in the interstellar medium near newly forming stars. This molecule can take part in the formose reaction, a chemical synthesis that itself can ultimately lead to sugars. Isopropyl cyanide, a compound that can be a precursor to amino acids, has been found in the interstellar medium. Formaldehyde (CH_2O) is also known to exist in the interstellar medium; this compound can take part in reactions thought to lead to compounds such as amino acids.

Complex molecules may be rarer and more difficult to find in interstellar clouds than simple compounds like HCN. As observational methods improve, we will doubtless find more of these complex molecules, but the general conclusion is clear: the interstellar medium contains many carbon-based molecules that can act as precursors and intermediates of chemical syntheses to construct living things.

Equally extraordinary and compelling evidence that the basic components of life can be produced in space can be found closer to home in meteorites. Although the concentrations of the over seventy amino acids found in carbonaceous varieties of meteorites are low (about ten to sixty parts per million), the compounds' presence nevertheless shows that the protoplanetary nebulae from which our Sun formed were propitious places for carbon chemistry.

So far there is no evidence that these amino acids have come together to form a simple protein chain. The conditions in the early Solar System seem to have been suitable for the formation of amino acids, but that is where complexity stopped. The more benign conditions of a watery environment on a planet's surface were needed if these building blocks were to become the chained, complex molecules we associate with life.

The amino acids in meteorites raise the obvious question of why we do not see amino acids strewn throughout interstellar space. Why the discrepancy? Maybe their low concentrations mean that they can be detected easily enough within a lump of rock in a lab, but dispersed in the interstellar medium, with all the other signals of chemistry out there, they remain elusive. Or maybe the discs of early solar systems make particularly favorable places for these types of compounds to form. A plethora of surfaces in proximity, temperature gradients, and volatiles such as water might make protoplanetary discs felicitous birth places for some of the more interesting reactions to bring forth the components of life.

Meteorites have given up their secrets as storehouses of much more than amino acids. Sugars, the building blocks of carbohydrates, and nucleobases, the letters of the genetic code, have been found, mixed in with a witch's brew of sulfonic acids, phosphonic acids, and other compounds. For each of the four major classes of biological molecules, or chains of compounds from which life is constructed— proteins, carbohydrates, nucleic acids, and membrane lipids—their monomers, or their basic building blocks, are found in meteorites.

Significantly, we do not find complex silicon-based compounds within meteorites, a storehouse of silicon variety that would make us wonder whether a battle might be played out between the possibilities of carbon- and silicon-based life wherever these materials land. The silicon compounds in meteorites are overwhelmingly dominated by the not-very-reactive silicate compounds. Meteorites show us the universality of complex carbon chemistry.

Meteorites come from asteroids and the rocky remnants of solar system formation, but an equally important class of objects, the comets, occupies the Oort Cloud, a spherically shaped reservoir of objects about 20,000 to 100,000 astronomical units away. Just beyond the orbit of Neptune, in the Kuiper Belt, yet another swathe of these icy bodies is to be found. Comets, like asteroids, are the leftovers of solar system formation. These objects are made up of a mixture of rock and ice, and the dark coloration of their nucleus is thought to be partly caused by organic compounds. In these objects, too, the same extraordinary story of carbon chemistry is being told.

By now, we have had a good chance to observe these frozen little worlds, using telescopes on Earth and in space, and by sending spacecraft to intercept and visit them. No mere blocks of ice, comets have been shown to contain carbon monoxide, carbon dioxide, and, from there on out, an increasingly complicated array of compounds, including methane, ethane, acetylene, formaldehyde, formic acid,

isocyanic acid, and, last and especially not least, the amino acid glycine, detected on comet 67P by the Rosetta spacecraft. The presence
of glycine raises some fundamental questions about whether other
amino acids and the other classes of molecules needed for life exist
on comets.

If these processes happened in our Solar System, unless there was
something very unusual about our location—and there is no reason
to suspect this—then they are happening everywhere. On the other
side of our galaxy, in Andromeda and millions of light-years from
here, amino acids, sugars, nucleobases, and fatty acids are raining
down on planets. Organic carbon chemistry is universal, and therefore, the likelihood that life, if it has taken hold elsewhere, would be
carbon based is high.

These stunning discoveries in space should not cause us to lose
sight of the possibility that our own world was the pot in which some
of these compounds for life were synthesized. The tendency of energetic environments to produce complex carbon compounds was
elegantly demonstrated in the 1950s by Stanley Miller and Harold
Urey. At a time when scientists lacked knowledge of the abundance
of carbon compounds in space and meteorites, the two researchers
set about to understand how chemistry might have made that momentous transition from prebiotic soup to replicating form. They
carried out a superb and simple experiment in the laboratory. Within
a container of gases—including methane, ammonia, and hydrogen—
they circulated water vapor. They added an electric discharge using
two electrodes to simulate early lightning on Earth, a bona fide Fran-
kenstein setup. After they energized the gases and added the water to
simulate the wet early Earth, what emerged was not a monster, but a
brownish goo that contained amino acids. Glycine, alanine, aspartic
acid, and many other compounds were synthesized. However, their
experiment used gases that we now think were not so abundant in
the early Earth's atmosphere. Change the composition of the starting
gases, and the yields and types of amino acids and other products are
altered. Nevertheless, the general idea that, given some energy, a few
simple starting gases, and some water, amino acids can be formed
was seminal in establishing that the organic chemistry of life is no

miracle. Indeed, to produce absolutely no complex organic chemistry when you add energy to a mixture of gases that contain some carbon atoms requires quite a bit of work.

The Miller-Urey experiment established that organic compounds can be formed on the surface of a young planet, not just in space. So here we have it—organic molecules forming in interstellar space and on the surface of a planet. From above and below, young planets are crucibles in which a selection of the enormous variety of organic compounds that exist can collect and become available for life.

The tendency for complicated organic chemistry to happen in places with some energy and basic ingredients is spectacularly demonstrated in all the organic chemistry occurring on Titan. Its methane lakes are the source of a cosmic factory of organic molecules. Its brown atmospheric haze is a product of atmospheric methane reacting with UV radiation high in its atmosphere, breaking down to form radicals that subsequently reform into ethane and complex chains of organic compounds, some of which drift in the upper atmosphere, producing the unusual hue.

Much of this material rains down on Titan's surface to produce vast deserts of organic compounds. In some regions of the moon, dunes made of complex carbon compounds ripple for hundreds of kilometers across the surface and up to a hundred meters high. If what is happening on the surface of this moon has been going on since the beginning of the Solar System, there would theoretically be a layer of ethane, the carbon molecule with the structure C_2H_6 and the precursor to many more-complex interesting molecules, about six hundred meters thick!

Are there building blocks of life mixed in with all this organic material on Titan? We do not currently know. Future robotic missions to Titan may answer this question, but whatever the answer, it does not impinge on the simple observation that on a moon with some methane and some radiation, vast dunes, lakes, and atmospheric hazes of organic compounds are being fabricated. Complex carbon chemistry is a planetary rash.

Some people think it uncanny that on Earth, conditions are just right for carbon chemistry to produce life. What we have seen suggests exactly the opposite. Everywhere, under conditions vastly different from those on Earth, carbon forms a huge array of versatile, reactive compounds. We find that, under a great range of temperature, pressure, and radiation conditions, of all the elements in the periodic table, carbon produces the greatest diversity of molecules. Yes, certainly, we find other compounds; even silicon-carbon bonds and silicon-nitrogen bonds detected in interstellar space suggest that extraterrestrial conditions can produce strange and intriguing new unions between elements rare on Earth. They give us pause for thought about the limitations of our chemical knowledge. However, glaring indefatigably through all these data is the array of carbon-based compounds that suggest that the variety of such structures on Earth is not unusual. What Earth may well have provided is the environment for these compounds to form long chains that then evolved into self-replicating entities. That step may require particular conditions not easily achieved in gas clouds and frozen dunes. But that event occurred in a universe where a carbon-based chemistry is the norm everywhere.

A more fascinating story develops when we expand our horizons beyond carbon to think about the other elements known to be spread through life. To build a life form, we need more than just carbon atoms. Of all the atoms available, five are ubiquitously attached to carbon to make more-complex arrangements: hydrogen, nitrogen, oxygen, phosphorus, and sulfur. These chemicals, along with carbon, are sometimes remembered with an unmemorable mnemonic, CHNOPS. Why these atoms? Is their ubiquity in life also a product of simple physical considerations?

These elements, too, just like carbon, owe their behavior to the Pauli exclusion principle. Hydrogen is always ready to bond with other atoms from which it can gain just a single electron to get a complete pair and fill its only electron shell. Throughout life, we find it bound to carbon. Hydrogen can be crudely regarded as a sort of mopper-up of single electrons that are available. That explains its appearance in everything biological.

The other four elements that, like nuts and bolts, hold living things together through their carbon networks—nitrogen, oxygen, phosphorus, and sulfur—intriguingly occupy a small quadrangle of the periodic table, huddled together near carbon. At the grand chemical scale, we can fathom this unusual affinity. All four have incomplete electron orbitals and will take part in bonding with other atoms to fill them. They inhabit a space in the periodic table where the size of the atoms is just right. Because the electrons can form bonds that do not need too much energy to break up, these elements are useful in the constant assembly and disassembly that characterizes the building and growth of living things. Nevertheless, the electrons in these atoms are attached tightly enough to make their links generally stable. In broad terms, these four elements are just very good at forming a diversity of compounds with multifarious properties. When they form bonds with carbon, they contribute to the imposing variety of chemical reactions required to be a successful reproducing and evolving life form.

On the face of it, nitrogen does not look like a very promising element for life. It makes an incredibly strong triple bond with another nitrogen atom to form nitrogen gas, N_2, which makes up a prodigious 78.1 percent of the Earth's atmosphere. However, once released from this chemical prison by catalysts in so-called nitrogen-fixing microbes or by nonbiological processes like lightning, the nitrogen is available to form a variety of useful bonds with carbon. One of its stable configurations is to sit between two carbon atoms and form the linchpin of the peptide bond, the linkage that holds individual amino acids together to form proteins. All amino acids contain nitrogen, which allows them to string together in this way. Nitrogen also turns out to be good at sitting between carbon atoms to make rings, and so it can be found in many of the prominent ring-shaped molecules in life, including the DNA base pairs. The nitrogen in these nucleic acids is the hookup to the sugars in the backbone, helping to hold the whole edifice of the genetic code together.

Let's now take a small step to the right in the periodic table and think about oxygen. This ubiquitous gas, so essential for animal life, sits neatly up against nitrogen in the table. Oxygen atoms perform a somewhat similar role to nitrogen, connecting carbon atoms in rings and forming attachments between carbon-containing molecules such as sugars, allowing long chains of sugar molecules, or carbohydrates, to form. Oxygen-containing sugars form part of the backbone of the crucial nucleic acids. We find oxygen in an array of organic molecules such as carboxylic acids, which participate in the synthesis of more-complex molecules such as proteins.

Phosphorus and sulfur, the two other molecules of our quad, sit below nitrogen and oxygen, respectively, and so they are larger with their extra electrons.

Familiar to most of us as the combustible material at the end of matches, phosphorus has insinuated itself into many of life's key molecules. Because it is larger than the other CHNOPS elements and its outer electrons form and break bonds more easily, phosphorus turns out to be a key component of many of life's energy-requiring reactions. Attach oxygen to it, and now you have a bond that can be broken to rapidly release energy in hydrolysis reactions where it is needed. The molecule ATP, with three phosphorus atoms strung together between oxygen atoms, has become one of the quintessential energy-storage molecules in all life on Earth, like a miniature battery for life.

The enormous versatility of phosphorus extends to the very structure of cells themselves. The phosphorus atom is found at the end of lipids, the long-chained carbon compounds that come together to make cell membranes. Even in the genetic code, this element is liberally distributed. Strung down the backbone of DNA, the phosphorus atoms form the links between the sugars, holding the structure together and helping provide longevity. The negatively charged oxygen atoms, found dangling from the phosphorus, give DNA a negative charge, which stops the molecule from leaving the cell since it is repelled by the negative charges on the inside of the lipid membrane. These negative charges have a second role: they help prevent DNA from being hydrolyzed, making the molecule much more stable.

To the right of phosphorus is sulfur, the biblical brimstone, or "burning stone," the yellow substance that adorns and drapes the vents and cauldrons of active volcanoes. Like phosphorus, its elemental association with fire and violence belies its beneficent use in life. The element is found in proteins. Two sulfur-containing amino acids in different parts of the long chain of amino acids that make up a protein can come together to form a disulfide bridge—two sulfur atoms linked side by side. These bridges help bolt together the three-dimensional structure of a protein as they link different parts of the chains into a well-defined shape that can carry out catalytic reactions in the cell.

These are just a few of the many uses of the four of the six CHNOPS elements necessary for life, but they illustrate the adaptability of these atoms and some of their different characteristics that find uses in cells as part of the machinery of life. Their deployment in many complex long-chained molecules shows just how useful they are.

Nitrogen, oxygen, phosphorus, and sulfur seem to have some convincing and useful applications in life. Evolution has used them often, accounting for their ubiquity and their elevated status to CHNOPS elements, but what about the alternative elements? Can we rule them out?

To the right of our chemically familial set of quads, and sitting next to oxygen, is fluorine. The element gained its fame in the process of fluoridation, a postwar attempt to reduce tooth decay by adding small quantities of fluoride to public water supplies. However, aside from this largely quiet entrance onto the stage of human society, fluorine is generally not used extensively by life. Its electron shell is almost full—with seven electrons in its outer electron layer, it would like eight electrons to make four pairs. Those seven are tightly bound to the nucleus, and the fluorine atom is highly desirous of that last electron in the set, like a small child who runs around in gleeful acquisitiveness to get the last color for a complete set of colored toys. This property makes it reactive, and when it binds to other atoms, it will not let go easily. The carbon-fluorine bond is the second-strongest bond in organic chemistry. Carbon-fluorine bonds are just too inert and strong to find much use in life.

But the fluorine atom is not a biological pariah. In the tropics, numerous plants and microbes use fluorine compounds as poisons to deter predators. As is the case for all the other non-CHNOPS elements throughout the periodic table, if some specific chemical property of a compound containing an element can find use in the struggle for survival, life will evolve to use it. The point is that the chemical properties of fluorine, accounted for by its electron structure, make it of limited use. It cannot be a universally used atom in life.

Underneath fluorine, the element chlorine suffers from similar problems. Although its greater number of electrons makes the atom larger and less desirous of the last electron, it still likes to get that final electron in the set, a propensity that gives the element sufficient reactivity to earn its place in the chlorine-based bleaches deployed in your bathroom to kill unwanted microbes. Not that it is discarded by life, either. It can be found in cells and carries out functions such as balancing the concentrations of different ions, but its chemical properties limit the scope of its use.

And beneath the nitrogen, oxygen, phosphorus, and sulfur? Below our quad are two intriguing elements, arsenic below phosphorus and selenium underneath sulfur. Both are found widely in living things and show us again that life never treats an element as an outcast. That said, the larger size of the arsenic and selenium atoms compared with their smaller CHNOPS counterparts makes their electrons less well bound; bonds with other atoms fall apart more easily. Nevertheless, this loose-binding property has not dampened enthusiasm for their possible role in alternative life forms.

Switch out phosphorus with arsenic below it on the periodic table, and you have invoked the weird and alien-like possibility of microbes that contain arsenic in their major molecules. In an article published in the journal *Science* in 2010, the astonishing claim was made that a bacterium isolated from the alkaline and arsenic-containing Mono Lake in California had replaced the phosphorus in its DNA with arsenic. Much ado resulted, launched with a press conference and wild enthusiasm for this new turn in the biochemistry of life. However, in a little less than a few days, the incredulity of the scientific community was being expressed online and in the press. Why? The tendency of arsenic-containing compounds to rapidly hydrolyze, or fall apart in water, because of that bigger atom, made it unlikely that the DNA molecule would hold together. The microbe, after further investigation, turned out to be a bacterium that uses phosphate in its DNA, just another member of the family tree of life that we know and love.

Some people might be tempted to say that this is how science works. Data are collected, claims are made, and sometimes they are refuted. This is how the scientific method advances the state of knowledge, however unpleasant the process can sometimes feel. However, even when this paper was published, the doubts about its credibility accounted for the rapid counterclaims that emerged. The estimated half-life of DNA bonds containing arsenate ions is about 0.06 seconds. If you replace arsenate with phosphate, the half-life jumps to about 30 million years. It would take very special circumstances or a lot of energy for a microbe to hold together a biochemistry with DNA in which the phosphorus atoms were replaced with arsenic, and this is true of many other phosphorus-containing molecules. The arsenic-using bacterium seemed unlikely from the start.

To put a positive spin on this episode, it shows that the elements that flank the CHNOPS elements have such similar chemical properties that it seems plausible that they could, under particular conditions, replace each other—so plausible that scientists could be enticed to think that a microbe could replace the phosphorus in its DNA with arsenic. Like all research, it stimulated more studies of arsenic in life. Out of any scientific claim, there is always the possibility of advances.

Nevertheless, despite this setback, we do know that life can contain arsenic. Its uses are enigmatic, but sugars containing arsenic have been found in some seaweeds, and arsenobetaine, an arsenic-containing molecule, is found in some fish, some algae, and even certain lobsters. Generally, though, arsenic is toxic to life. Its great propensity to share electrons means that it interferes with other molecules, reacting with them and disrupting metabolism. Many life forms have pathways to reduce or remove the toxic effects of the element.

Arsenic's next-door neighbor to its right, selenium, could plausibly replace sulfur, the element above it. This prediction is borne out in life. The element is found in one of life's stranger amino acids. The so-called twenty-first amino acid of life, selenocysteine, is diligently included in some proteins. The energetic cost and modification to the genetic code required to include this amino acid means that its incorporation into living things is not a mere fluke of an exchange with sulfur. Selenium must perform a vital task. Some proteins in which it is found, such as glutathione reductases, prevent damage caused by oxygen radicals, which are reactive and potentially biologically damaging states of the oxygen atom. As an atom larger than sulfur, selenium can hand over its electrons more easily. This property plays a role in its ability to neutralize, if you will, the damaging free electrons in oxygen radicals. Once the selenium atoms have carried out this important function, they are more easily returned to their original state to carry out similar reactions. This reversibility, again because selenium can gain and lose electrons more easily than sulfur can, makes selenium useful in these roles. Moreover, with selenium, proteins seem to develop more resistance to oxidation, the process of losing electrons, caused by various types of chemical attack.

Again, we see the same pattern: arsenic and selenium are not entirely rejected by life, but their size and the behavior of their electrons put them just outside that ability to be useful in many contexts. They have a specialist use in a few situations.

To complete this tour around the fringes of the CHNOPS elements, let us examine one element near carbon in the periodic table—an element that we have so far neglected. Boron, an interesting little atom with three electrons in its outer orbitals, can share those electrons with other elements. It can form bonds with nitrogen and, in so doing, generates compounds like borazine, which is analogous to the ringed carbon compound benzene. Boron lacks the chemical versatility of carbon, but like the other elements that surround the CHNOPS elements in the periodic table, it finds biological uses. It is an essential trace element in many plants, microbes, and animals and is thought to stabilize cell membranes and transport sugars. This is no ephemeral role: boron deficiency is one of the major trace-element deficiencies in agriculture, causing crop failures in plants from apples to cabbages. Our knowledge of the diverse roles of boron in biology is still in its infancy.

We are left with a general picture of what the Pauli exclusion principle does to shape life. There is a core of elements whose electron structures are sufficiently good to create stable bonds, but which can nevertheless be broken with sufficient ease to generate the huge assortment of compounds useful to life. These core elements occupy a little quintet in the periodic table—carbon, nitrogen, oxygen, phosphorus and sulfur, with little hydrogen ubiquitously hanging on whenever there is a spare electron going. These elements have just the right atomic size and the right number of spare electrons to allow for binding to each other and to some other elements to produce a molecular soup sufficient to build a self-replicating system.

Surrounding the quintet are elements with similar chemical properties, but in which the atomic size and number of electrons make them either a little too unstable or a little too reactive to be good at producing that fine balance between stability and reactivity for making the molecules of life. Although they are not ideal for a wide miscellany of jobs, they find use when their characteristics make their chemistry just right for a particular purpose.

Scattered through the remainder of the periodic table are the rest of the elements that, to lesser or greater degrees, find application in many places where their electrons can do something useful. Iron takes center stage in the choreography of gathering energy for life as it shifts electrons around, the core process of snatching energy from the environment with which to grow and reproduce. Elements like vanadium and molybdenum, with their dexterous electrons, pop up here and there, such as in cofactors in proteins that help speed up reactions.

From sodium to zinc, many elements not among CHNOPS, particularly the metals, are much better at forming salts, such as common salt, sodium chloride. Although these atoms can form large macroscopic structures, as a large lump of table salt will testify, they are hugely regular and highly monotonous repeating units of atoms that seem poorly suited for producing living things. These salt-making habits explain why tin and lead Horta seem unlikely. The imaginative may well raise their hands in disdain and point out that if we add defects and faults into salt crystals, with some other elements thrown in for good measure, could we not get complexity suitable for life—self-replicating, evolving crystals?

The conditions on Earth are just right for a vast collection of crystals and salts. Yet despite 4.5 billion years of chemical experimentation, the self-replicating, evolving crystal has yet to be reported. No good scientist would use this evidence to conclude that the phenomenon was impossible. Mineral surfaces as places for the assembly of early life have played an important part in ideas about the construction of the first self-replicating organic compounds, but many elements in the periodic table appear better suited for forming rather mundane salts in the natural environment. Life evolves to use these elements, for carrying out tasks in electron shuffling and transport, but in themselves, these elements do not seem sufficiently flexible in their bonding patterns to produce an abundance of complex molecules from which we can build a living system.

The process of evolution rummages through the periodic table, testing elements and selecting those whose electron arrangements facilitate chemical reactions that make an organism better able to survive and reproduce.

However, an enduring question remains: could life be something other than carbon based? Is the chemical structure of life universal? I think this question is loaded. We can say that life on Earth is carbon based if we want to categorize life according to the quality that carbon is a dominant element in the assembly of its molecules. However, life is obviously periodic table based. In some environments, organisms make use of selenium more than others do; in other circumstances, even fluorine is found in life, though this element makes no appearance elsewhere. Life is not somehow transfixed by carbon—living cells use what they can. The only principle at play here is that the elements used to construct a replicating, evolving system must have arrangements of electrons capable of yielding the chemical reactions and bonds that allow for the integrity of the living system and its continuity. In the conditions found on Earth and in many other environments in the known universe, carbon, with its binding properties to hydrogen, nitrogen, oxygen, phosphorus, and sulfur, can assemble the basic chassis of a replicating, evolving system. Other elements provide the fine-tuning to make that system work better than it might otherwise, to produce a full panoply of diverse molecules. A bountiful potpourri of elements has been (and is being) tested continuously by the process of natural selection for its uses, but no known cell has yet shown a strong selective bias to replace most of its carbon molecules with something else.

Given what we know about the behavior of many elements at low and high temperatures, at different acidities and pressures, and at other extremes, it seems doubtful that under other physical conditions, other elements will step into the shoes of carbon and replace it with the combination of chemistry useful for a living entity.

Ultimately, no physical conditions, at least conditions under which the atoms of the periodic table are stable, will alter the basic electron configurations of the atoms. The rate at which they undergo reactions and how they interact with other atoms will be modified by their environment. However, their core characteristics, set by the Pauli exclusion principle and its implications for electron stacking, are invariant. I would expect that on any planet in the universe, given the limited range of elements available in the periodic table, life would forage through this table in the blind process of evolution, just as it is doing on Earth, and arrive at the same set of elements. Yes indeed, the use of the elements, their applications, and their abundances in living systems will show great variance, just as they do on our planet, but the fundamental role of the major elemental players in life is likely to be the same in every galaxy in the universe.

The idea that physical principles explain why carbon is the central atom in life's construction and water the milieu in which life operates brings me to one final and essential point. There are two possible directions for our conviction that carbon- and water-based life is universal, that physical principles narrow the possible chemical architecture of life. The first I will call the soft view. In this view, alien biochemistries, be they silicon-based life forms in liquid nitrogen or acid-resistant life in sulfuric acid clouds, are possible but rare. The unusual conditions required for these alien life forms and the relative cosmic abundance of carbon chemistry and water make them unlikely types of life. We might also call this the abundance-based view of carbon-water bias.

The second view is the hard view, what we might call the chemistry-based view of carbon-water bias. According to this view, no other forms of life are possible and the chemical properties of other elements such as silicon or alternative solvents such as ammonia are inadequate in their diversity to drive the formation of life, regardless of their cosmic abundance.

My view on life is minimally the soft view and leaning strongly on the hard view. The abundance of carbon compounds and water makes it likely that life elsewhere, if it comes into existence, is carbon and water based. As we have seen, the sheer universal distribution of carbon chemistry and water and the propensity of these two substances to collect on planetary bodies of all kinds suggests that they provide the most likely ingredients for life.

However, why do I only "lean strongly" on the hard view, the view that carbon- and water-based life is the only possibility? Simply because it would be poor science to be dogmatic.

Can we rule out that in some planetary system somewhere, there is a body on which ammonia oceans have formed because of an unusually high abundance of this liquid in that area of space, where the geochemical conditions have allowed for self-replicating simple organisms to emerge? Can we rule out that in some planetary crust somewhere, a low abundance of oxygen or the right physical and chemical conditions have frustrated the formation of silicates and in some little pocket, a self-replicating set of silicon compounds has come into existence?

In the absence of much greater knowledge of the chemistry of the elements, it would be foolish to dismiss these possibilities, even though a dogmatic support for hard carbon-water bias is rhetorically appealing. Our still-incomplete understanding of both chemistry and the full diversity of conditions under which planetary systems form requires that we keep an open mind. Nevertheless, even soft carbon-water bias leaves us with a view of something universal about life on Earth, the simple physical principles that nudge and narrow the atomic structure of living things.

CHAPTER 11

UNIVERSAL BIOLOGY?

WHATEVER OBSERVATIONS ABOUT THE preponderance of carbon-based chemistry and water in the universe might entice us into drawing strong conclusions about the structure of life and its potential atomic similarity elsewhere, there remains an inescapable fact, an ineluctable limit to our current knowledge—we have only one planet from which to draw our conclusions. Using terrestrial life to draw inferences about whether universal physical principles operating on life would result in similar or identical outcomes, if indeed life has occurred elsewhere, necessarily pens us in. Sometimes this is referred to as the $N = 1$ problem. Any scientist with even the most meager self-respect feels uncomfortable drawing conclusions from a sample size of one. For this reason, many people feel that a discussion about the extent to which the characteristics of life are universal or inevitable is a flawed enterprise.

The effort to find what is common in all life on Earth and, hypothetically, elsewhere is sometimes referred to as an attempt to discover if there is a universal biology. It is easy to get trapped into an argument about whether biology is autonomous and has laws distinctive to physics, but naming conventions are uninteresting. The term *universal biology* is simply shorthand for asking "What aspects of a self-replicating lump of matter that evolves in response to its environment are common to all examples of such material?" I have assumed that self-replicating matter that evolves is my general working definition of *life*.

We create a problem for ourselves when we sharply define *biology* and *physics* as two separate areas of science since we are led, by our own mental slavery to human language, to start asking meaningless questions about whether some "laws of biology" are universal. In truth, there are no laws of biology or even physics; there are only laws that determine how the universe works. Those laws are inseparably operational on all forms of matter. Calling ourselves physicists or biologists is an unfortunate tribalism. Biologists just happen to focus their efforts on particular lumps of matter that do some interesting things, and we often call those things "life," but biologists and physicists share a common interest in matter in the same universe and in the potentially universal principles they can draw about the ordering of this universe.

Although many people think that the $N = 1$ problem makes the question of what characteristics of biology might be universal impossible to answer, we are probably on less shaky ground than we might assume. Many of the characteristics of life that we have explored in this book seem aligned to physical rules that we infer would work on any life anywhere.

We have seen the reasons for carbon as a major element of life and the preponderance of complex carbon chemistry through the universe over other elements. Water seems like a very good candidate for a universal biological solvent, in light of its physical properties and its abundance across the cosmos. But we have seen other contenders for universal features of life. The tendency of protein chains to fold together toward their lowest energy state results in a few predictable folding types. We might predict, from these observations, that any chain of compounds that makes up a life form would fold together in a limited number of ways in which universal thermodynamic considerations at least play some part. Cellularity, the physical process of concentrating molecules against environments that naturally tend to disperse and diffuse, is a good contender for a universal characteristic of self-replicating systems of matter.

At the scale of whole organisms, evolution has explored many millions of experiments in form, not merely one. Convergent evolution offers us a vast menagerie of animals whose shapes and structures have been molded independently into similar forms by physical principles, such as our friends the moles. The scaling laws that define the interrelationship between animal size and properties as diverse as metabolic rate and life span across a vast range of creatures, from cats to whales, suggest principles applicable to any type of life. These observations show us that within our one biosphere, we have a collection of experiments from which to examine how physical principles drive common forms of life at its different scales. A single biosphere does not mean we cannot fathom general principles on how evolutionary products are fashioned by physical laws, with the implication that we might use these principles to predict the nature of life elsewhere. The argument that because all life on Earth shares, as a common heritage, a so-called last universal common ancestor, any discussion about universal biology is flawed certainly does caution us to be careful about similarities caused by common biochemical and developmental heritage. But even with this caveat, we can still discern why the successful origin of life that led to our particular evolutionary experiment has the atomic characteristics it has. We can still observe organisms converging into forms that reflect physical principles, even given certain ancestral similarities and channels of developmental biology.

However, we often find it less easy to discover where contingency might have a powerful role to play. We might wonder about the genetic code. That the code employs particular bases and is optimal with four may well be a predictable outcome of using the class of molecules with which we are familiar. Although our genetic code looks much less of a freak accident than we once thought, we are still left with the question of how many radically alternative chemical systems of hereditary information could exist in living things. Is it zero, ten, a hundred, or some other number? Other questions about our genetic code abound. Do we need an intermediary, a messenger such as RNA, between the code and functional molecules?

Even if we cannot define exact molecular combinations that are universal, we might be able to say something about the expected universal features of their general chemistry. Perhaps, like DNA with its negatively charged phosphate backbone, we can at least predict that the molecular stuff of other life would be made of long chains with repeated negative or positive charges along its length. Synthetic biology and our continued adventures in chemistry may well allow us to answer these questions, eventually enabling us to further describe the characteristics universal to life.

Perhaps it is a mistake to focus on particular structures of life, but instead we should direct our attention to life processes or products that we can show are inextricably linked to inescapable physical principles. These processes and products may be a more likely path to discover what is universal about life.

For living things to reproduce and evolve, generating the panoply of creatures we observe on the planet, they must contain a code that passes from one generation to the next and contains within it the information needed to produce new living things. That code must not be perfectly reproduced, or there will be no variation on which the environment can act to generate new forms. However, the reproduction of the code must not be so imperfect that each generation contains many errors, causing life to degenerate into an amorphous organic mess caused by an "error catastrophe." Perhaps this feature of life, which inextricably links it to the evolutionary process, is universal. That life is a "system of reproducing matter containing a code whose reproduction lies between perfection and error catastrophe" might be a contender for a universal physical characteristic of matter that makes life. Here, then, we might seek universality in life by identifying those characteristics required for matter to undergo evolution, one of features we have decided circumscribes the matter we call life.

That life is a system of matter that dissipates energy and uses energy to reproduce and evolve leads us to look toward its energetics and thermodynamics as a potential source of defining universal characteristics. For example, if we think that electron transfer might be a universal way to gather energy, then we might single out a few elements or molecules that any living thing can use to gather energy.

We know that life could use hydrogen and carbon dioxide as a source of energy (and make methane in the process) anywhere in the universe with a suitable biochemical machinery because these two compounds produce a thermodynamically favorable, energy-yielding reaction in many environments. This is a fact set by physical laws, not by contingent evolution. Thus, tabulating the universal electron donors and acceptors that life can use to gather free energy from the environment is a rather trivial matter, and this exercise allows for universal observations about the energetic potentialities of life.

Energetic considerations might yield other predictions. If alien life forms eat each other, we could also predict food chains with fewer larger predators at the top of these links, where energy is limited, and a larger number of small creatures in the lower levels of the food chain, as we see in life on Earth. From the electron transfer chains at the subatomic scale to food webs at the population scale, these facets of life are driven by thermodynamics. Moreover, an exploration of the biological consequences of laws that instantiate the movement of energy, such as the basic laws of thermodynamics, might well allow us to define the universal forms of life.

When it comes to predicting the forms of living things, we might look at inviolable equations such as $P = F/A$ and explore likely universal outcomes of the imposition of this equation on organic forms—moles and wormlike entities at the scale of whole organisms. Many of the equations we have explored in this book, unexceptional and universal in their applicability, provide a framework for predicting the assembly and general convergence in the architecture of living things.

Although there seems to me to be hope of making progress in exploring the potentially universal processes and structures of life, it is probably facile, but worth saying, that the most reliable way to investigate the scope of contingency in how physics limited life on Earth would be to examine another example of life. What are our prospects of ever finding another example of life outside our planet?

We do not know whether we will find outside our planet life that is truly an independent genesis, the evolutionary experiment run from scratch. In our own Solar System, we have found numerous watery environments, such as ancient Martian terrains that seem to have been habitable, potentially supportive of life. Perhaps there is life below the surface of that planet today. Substantial liquid water oceans exist in the icy moons of the outer Solar System, including Jupiter's moon Europa and Saturn's moons Enceladus and Titan. Do they house independent experiments in evolution? Even if they do, a confounding problem could be that their biota might be related to life on Earth, existing as it would within our Solar System, where rocks, and potentially their hitchhiking life, have liberally been shared between the planets in the form of meteorites since the planets first coalesced from the early protoplanetary disc. Nevertheless, the search for life in our Solar System is a worthy scientific goal because if we did find life independently evolved from Earth, we could assess the universality of biology. If we find life related to Earth's biota, or no life at all, we will learn less about biology, but we will have learned something about the distribution of life and its capacity or lack thereof to originate or be transferred within a solar system.

Besides the impressive strides in the robotic exploration of our own Solar System, extraordinary advances have been made in finding Earth-like worlds around other stars. Do these breakthroughs offer us any hope for answering our question of what features of life may be universal? It will be a long time, if ever, before we visit exoplanets orbiting distant stars and can sample their biospheres. Even planets a few to tens of light-years away would be multigeneration missions with the best conceivable propulsion systems we could build at the current time. It is premature to declare that the discovery of exoplanets opens up the real possibility of expanding the set of biospheres with which to explore life's universality.

Nevertheless, retreating from the heady ambition of finding another biosphere to test some of the observations expressed in this book (although maintaining our hope that this might one day happen), there is something to be said for delving into a discussion of these quite remarkable exoplanetary discoveries. At the very least, in a more narrow-minded way, we might ask some slightly different questions. How alien are these other worlds? Even taking as our hypothetical starting point the life we know on Earth, would we expect the environments of these planets to channel evolution in alternative ways?

Although this may seem a little speculative compared with the more data-rich basis of our previous intellectual escapades, such thinking can sometimes illuminate the view of our home world, stimulating us to ask fresh questions about the forces that shape the products of evolution on Earth. Using other planets to see the Earth in a fresh light is often a fruitful way to open up the human mind. So in the interest of expanding our minds to the full scope of the question of whether the structure of life is universal, let us explore briefly this new horizon on alien worlds and indulge ourselves in some evolutionary thought.

When eager scientists turned their gaze to distant stars to find planets, they expected to find these worlds in planetary systems much like our own. In the outer reaches beyond a star, giant gaseous planets like Jupiter and Saturn would orbit, the frigid temperatures from the early dawn of those solar systems, like ours, favoring the condensation of hydrogen, helium, and other light gases that in the inner regions near the star tend to be vaporized into space. Closer to the star, where those gases disappear, are left rocky remnants, the collided and congealed fragments of material that built the so-called terrestrial planets, like Venus, Mercury, Mars and our own special home, Earth. Small bits of rock near a star, big balls of gas far away. That seems to work nicely.

It was something of a surprise then that when astronomers did detect another world orbiting a distant star for the first time, it turned out be a giant of a world, orbiting once around its sun in a preposterously short five days! So close was this planet to its star that astronomers had no choice but to call this new object a hot Jupiter. That discovery of this first so-called exoplanet in 1990, orbiting around the science-fiction-sounding star Pegasi 51 and 50.9 light-years away from us, threw astronomers into a certain confusion. How could it be orbiting so close? Such a large gaseous world must surely be vaporized? The discovery forced astronomers, early in the quest for these distant planets, to revise their models of how solar systems formed. Only one theory could properly account for these giant worlds so tightly hugging their parent stars: they had migrated from the outer reaches of their solar systems inward and taken up residence in neighborhoods thought to be exclusively owned by small rocky worlds. This discovery was the first inkling that our own Solar System may not be typical.

Pegasi 51b was the first in a flood of many peculiar types and orbits of planets that would litter the scientific literature as novel discoveries piled up. About twenty years after the first observation, the surprises keep coming, but several things have become apparent and will long endure. Our Solar System's architecture is not typical. In other solar systems, planets are strung together in many configurations, with migratory planets throwing a wrench in the works and leading to a host of systems with small and large planets in different positions depending on how gravity arranged the alignment of worlds. Many planets do not occupy the nice near-circular orbits that are a characteristic of our Solar System's major planets. Planets can take on wild elliptical orbits, tracing arcs that take them far away from their stars and hurtling inward in close shaves, a little like the comets that we are more familiar with. These extreme trajectories are a consequence of the gravitational perturbations and interactions that shaped these systems in their early years as new planets formed and others migrated. Some pathways are so extreme that planets can even be hurled out of their star systems entirely, exiled into the infinite abyss.

What exoplanet hunters around the world have found is nothing less than a feast of bizarre and wonderful places. Alongside the bounty of hot Jupiters, astronomers have found hot Neptunes, slightly smaller gas worlds the same size as Uranus and Neptune.

These were by no means the only oddities. Some gas giants orbit so close to their star that it is thought that the intense heat of the star causes the gases to expand into giant envelopes, inflating the atmosphere. These worlds, or *puffy planets*, have a very low density. One such object, HAT-P-1b, just larger than Jupiter, lies 453 light-years away and orbits its planet every 4.47 days. It has one-quarter the density of liquid water. This inflated ball epitomizes worlds that would have been unimaginable to astronomers just a few decades ago.

What interests us most are worlds where the conditions for life might exist. As exoplanet detection methods have improved, the size range of planets capable of being detected has been reduced to smaller and smaller spheres. Somewhere between the mass of the Earth and Neptune-like worlds, there is a gray region where planets make the transition between being giant gas worlds or small rocky worlds. Between these two extremes are the planets sometimes called *super-Earths*. The term is somewhat of a misnomer because these places are not necessarily Earth-like. Many of them are likely to be uninhabitable or might have originated in conditions very different from our own planet. Some are likely to be ocean worlds, where a good fraction of the mass is water.

Crucial to the quest to find truly Earth-like planets is to find planets in the habitable zone around a star, where the solar radiation is just high enough to allow liquid water to persist on their surfaces. Get too close, and you end up like Venus, the oceans boiled away, helped along by a runaway greenhouse effect where the dense carbon dioxide in the atmosphere traps heat, causing all the water to evaporate. Form a planet too far away, and it generally remains a frigid, wintry world. The habitable zone, sometimes called the *Goldilocks zone*, is the annulus where temperatures are just right, the zone in which our own evolutionary experiment has explored its capacities.

It was a planet orbiting Kepler 452, about fourteen hundred light-years away, which took the first title of being a genuinely Earth-like world. Although it is slightly larger with a diameter about 60 percent greater than Earth's, it was the first Earth-like world to be found orbiting a Sun-like star and within the habitable zone. It is slightly more aged than our own world—it is about six billion years old.

The sheer quantity of data gathered by instruments such as the Kepler space telescope has been used to suggest that about 5 to 7 percent of Sun-like stars in our galaxy might host Earth-sized planets in the habitable zone. Take these numbers and scale them up to the whole galaxy, and you are confronted by outrageous numbers, guesses that the Milky Way contains something like eight billion Earth-sized, potentially habitable worlds! Well, we can argue about whether that should be five billion or ten billion, but this is nitpicking.

The discoveries of exoplanets have revealed just how common small rocky worlds are.

How did astronomers find these planets? The story of exoplanet discovery has filled the pages of entire books, and although this story is somewhat tangential to our focus on life and physical principles, it is well worth a brief detour. The discovery shows the fascinating connections now being made between researchers in the physical sciences and those in the biological sciences; both groups are motivated by an interest in understanding the physical conditions on other planets, conditions that we assume would influence any life.

Two decades ago, astronomers and biologists had very little to say to one another (apart from general coffee-room chatter). However, the methods that astronomers are using to look for exoplanets are revealing Earth-mass planets that we might eventually examine for signs of life. In this endeavor, physical and biological sciences come together, and the forging of this alliance may well yield a rich harvest of ideas about how astrophysical and physical principles shape the conditions for planetary formation, the surface characteristics of planets, and, therefore, the potential physical environments in which life might emerge. This is one of the most exciting emerging areas of astrobiology.

It is not immediately obvious how to detect a planet around the blazing inferno of a star. The light reflected from even the grandest Jupiter-like objects is many billions of times less than that generated by the immense burning fusion reactor of a star. To overcome these limitations requires some ingenuity. And astronomers have brilliance in no short supply.

Picture yourself with a telescope staring at a distant solar system edgeways. Caught in the lens of your telescope is the bright star that lies at the center of the solar system, but then the star dims. Gracefully passing in front of it is a planet in orbit, its opaque form blocking out some of that starlight for as long as it passes in front. That dimming is not much, maybe 1 percent of the light or less, but with a good telescope and some equipment to measure the light, you can see the temporary drop in intensity as the planet passes by. This method, the transit method, has its limits—the solar system must be observable edgeways. It is no good observing such a system from the top down since as the planet orbits the star, it will never intersect with the star's light. Nevertheless, the transit method has been extraordinarily successful. The Kepler space telescope, named for seventeenth century astronomer Johannes Kepler, who first elaborated the laws that define the orbits of planets, used this method to find over a thousand planets. The limitation of needing solar systems that are oriented edgeways to us has not hindered scientists from finding a bounty of planets.

Astronomers have found some other clever ways to find planets. You can detect these worlds by observing the wobble they cause in the star around which they orbit. Imagine or recall yourself at a wedding, watching couples dancing at a reception. A couple grasp each other's hands and spin wildly around. They spin around their common center of mass, around an imaginary point that lies somewhere between them, and both become dizzy at the end. As they leave the floor, the unfortunate young woman is intercepted by her large uncle, who also demands a dance. The petite woman holds his hands, but now this time, his mass is so large she flies around him, his position almost fixed in the dancefloor. However, he is not entirely fixed. The common center of mass between them is an imaginary point that lies around the edge of the gargantuan uncle or even within him. He too wobbles and spins around the common center of mass, relatively less affected than she is, with her small and light frame that spins around his bulk. When they have finished, she is positively giddy, and he too lurches and stumbles back to his table.

So too in the world of astrophysics. A planet does not strictly orbit its star. Both objects orbit around their common center of mass. However, a star is so massive that the common center of gravity is essentially buried in the star; it hardly notices the tiny planet tugging on it as the smaller body dances around the star. Although the effect is small, it is still there and the star subtly wobbles around like our slightly unstable uncle, its spin influenced by the mass of the planet. If there are many planets in orbit, the movements will be more complicated, the speed and pattern of the lurching revealing many worlds.

How can we, many hundreds or thousands of light-years away, see this almost imperceptible change? The good news is that this is where ice cream takes center stage. As I'm sure happens to you every summer, the mere sound of the tune from an ice cream truck causes you to rush out into the street in happy exultation. Now as the truck approaches, the pitch of its melody is high, but as the vehicle passes you (your disgust at its failure to stop overwhelmed only by your fascination with physics), the pitch gets lower. You can observe the same effect with the more sobering sound of an ambulance.

The Doppler effect, named for Christian Doppler, an Austrian physicist who proposed how it worked in 1842, causes this alteration in pitch. The effect is easy to understand. As the ice cream truck approaches you, it is emitting sound, but since the vehicle is moving, each successive soundwave is produced closer to you than the previous one, hence the time between the waves is reduced (they become higher frequency, which is perceived by you as the higher pitch of the sound). When the rascal driver disappears into the distance, each successive wave of sound emitted is further away, since the truck is traveling into the distance, making the frequency of sound lower; the sound waves are essentially stretched.

What does ice cream have to do with exoplanets? It transpires that the Doppler effect also influences light, because, like sound, light moves in waves. If a glowing object like a star is moving toward you, its light will be ever-so-slightly bluer than if the star were stationary, since as it comes at you, those wavelengths of light are slightly squashed to shorter wavelengths, in the blue region of the spectrum. Similarly, as the star travels away, then the light is slightly redder, because by stretching light, the wavelengths become longer.

As the giant star wobbles around the common center of mass with the planet that orbits it, seen sideways from Earth, the star very slightly seems to approach and recede. This wobble introduces into its spectrum a subtle shifting of the light, bluer as the star comes toward you, red as is recedes. By accurately observing the changing spectrum of light, this Doppler shift method of detecting exoplanets, sometimes also called the *radial velocity method*, can be used to ascertain the mass of orbiting planets. Combine these data with information from the transit method, which also gives you the size of the planet, and we can work out the density of a planet, which is important if we are to establish its composition.

With all these bizarre new worlds, it would be easy to return to H. G. Wells's speculations about silicon beings walking along the shores of liquid iron oceans. We could declare that now that we have found an apparently endless variation in forms and sizes of worlds, we must also be open to the same possibilities in biology. However, despite the plethora of physical features we might find on these planets—features that might suggest strange environments for alien life— we can bring to bear our discussions in this book on our assessment of how physical principles might shape any life on them.

From observing the spectra of light, we know that all exoplanets are made of the same elements in the periodic table. The knowledge that exoplanets are made from the same periodic table, an extraordinarily trivial point of fact (no one was expecting anything else), leads to many straightforward and immediate points. The same restrictions that make carbon better for all those complex molecules we observe in the interstellar medium, on Earth and in life, characteristics with their roots in the quantum world, apply on exoplanets as well. On any rocky world, we would expect silicon to be primarily in minerals and would expect carbon to be the preferred choice for building self-replicating, evolving entities. Thus, at the atomic scale, we are in a position to make some predictions about the potential structure of life on other planets, if it exists.

The fact that water is a common molecule in the universe means that it is likely to be the most common liquid to slosh around on the surface of rocky worlds around distant stars. It may be mixed in with other compounds such as ammonia and, under enough pressure, liquid carbon dioxide, but it is an abundant solvent likely to stare any would-be biological experiment in the face from the moment of its inception.

Even the energy that other living beings might gather from the free energy in electrons must be the same sort of energy sources we observe on Earth. The electron donors and acceptors in the periodic table or in the compounds made from universal elements are the same anywhere across the universe. They are not an idiosyncrasy of Earth; they are mandated by the thermodynamics of the elements and compounds made from them. Temperature and pressure conditions will alter the thermodynamic plausibility of given reactions, and the abundance of different compounds will change what sources of energy are available to life, just as these variations occur across Earth. We cannot necessarily predict which types of energy sources will be more favorable or prevalent on and within a given exoplanet without some substantial knowledge of its geochemistry and geophysics. However, the reactions available from the tool kit of the periodic table will be the same.

Within these restricted boundaries, planets might certainly host more or less bounteous biospheres than Earth. Some features may make other Earth-like worlds more fecund for life. René Heller from McMaster University in Canada and John Armstrong from Weber State University in Utah speculated on what it would take to make a planet more habitable than Earth. They imagined many things that might make a "superhabitable" planet even better for life than our own verdant oasis. Planets slightly larger than Earth might accommodate more biomass or even more diversity. Ecologists know well that the more land or continental shelf there is, the more diversity of life you can fit in. Planets with more water bodies in the interior of continents and fewer arid areas might support greater biomass. Under the light of stars with lower levels of UV radiation, planets might be less hostile on their surfaces, where radiation damage is less of a concern for exposed life. We can imagine how differences in planetary size, the ratio of land to ocean, the surface temperature, or atmospheric composition would alter the conditions for life.

However, even throughout the history of Earth, there have been large transitions that have influenced the diversity and biomass of the planet. Continents have shifted, the atmospheric composition has changed, and continents have become less or more arid. Astronomical events, such as asteroid and comet impacts, have periodically harassed our planet. Some of these events have been so profound that they have caused mass extinctions. Defined in terms of animal and plant biomass, Earth today is more habitable than it was five hundred million years ago. Yet none of these alterations to planetary conditions have changed the restrictive influence of physics on life.

Am I saying that life on these exoplanets would be tediously like that on Earth? Not exactly. Even though the scope of the evolutionary experiment, restricted by physics, would obviate the wild forms of life that Wells imagined, there is still room for great variety to be produced on worlds where different physical conditions prevail.

To explore how different physical conditions elsewhere might shape life, but nevertheless allow for fascinating diversity, it is enlightening, even entertaining, to speculate briefly on some possibilities. One physical factor exerts itself on life across the whole of Earth and would do so on any other planet in the universe. Even Darwin, in the final paragraph of his book, cared to mention this factor: gravity. On many rocky exoplanets, gravity would be different from this phenomenon on Earth. How would this single factor influence life? We can work out the surface gravitational acceleration (g) on any world using the simple relationship that the acceleration of gravity is proportional to the mass of a planet and inversely proportional to its radius squared, so

$$g = GM/r^2$$

where G is the gravitational constant (6.67×10^{-11} m^{3}kg^{-1}s^{-2}) and M is the mass of the planet.

Consider an exoplanet ten times the diameter of Earth. The mass of the planet is related to its volume, given by $(4/3)\pi r^3$. The mass therefore scales as the radius is cubed. To get the gravity, we divide the mass by r^2, showing simply that the gravity scales as $r^3/r^2 = r$. Assuming for simplicity that the bulk density of the planet is much the same as it is for Earth, then on this planet with ten times the diameter of Earth, its gravity will be ten times greater.

Let us consider a large cow-like creature wandering the fields of this distant planet. All other factors being equal, the weight (given by mg, the mass of the organism multiplied by the gravitational acceleration) will be ten times greater. The weight of the animal must be supported by the cross-section of the legs, and if the weight is increased by ten, then by increasing the diameter of the legs by 3.2 times, the cross-sectional area of the legs is increased by ten times. This increased diameter will produce the same downward force per cross-sectional area of a leg as on an Earth cow's leg cross-section.

The alien cow must have thicker legs than our familiar cows and maybe a smaller body to compensate. Although stronger bones or muscles, evolved under a higher gravity, might yield some organisms not much different from life on Earth, large life forms on this high-gravity planet would probably bear the anatomical imprint of this higher-gravity environment.

And what about our alien fish? You will recall that the force acting in a fluid is given by $mg - \rho Vg$. The first term gives the weight of the animal. On our exoplanet with ten times the gravity, the weight of our fish is therefore ten times greater. However, because the weight of the water being displaced by the fish also scales with the gravity, the buoyancy term, ρVg, which contains g, is ten times greater on this new world. Increase the gravity by ten times, and there is no net effect on the fish. Fish and whales care little whether they are on Earth or a super-Earth.

Diminutive creatures would be even less affected by the difference in gravity. At the scale of a ladybug, gravitational forces become almost irrelevant. As we learned, molecular forces are what dominate the ladybug's world. Even the forces of attraction in a thin layer of water under its feet are sufficient to keep it attached to a vertical wall, competing against gravity that would otherwise yank the bug off. Ladybugs, however, are not immune to the force of gravity. When a ladybug closes its wing cases in mid-flight, it will fall to the ground as surely as will a person leaping off a wall, but the resistance from the atmosphere is more effective in limiting the speed of the descent of such a light-weighted creature.

We might think about the effects of gravity on creatures jumping off alien cliffs or from the branches of tree-like plants. When an object falls, it reaches a terminal velocity, which is the fastest speed it will go. It is a trade-off between gravity pulling down and the drag of the air or fluid that prevents it going any faster. We can work out this maximum speed (V_t) using an equation:

$$V_t = \sqrt{(2mg/\rho AC_d)}$$

where m is the mass of the object, g is the gravitational acceleration, ρ is the density of the air or fluid through which the object drops, A is the surface area of the object, and C_d is the drag coefficient of the object, which expresses the extent to which drag or resistance will slow it down.

You will notice that the equation involves the mass of the creature, which is important, as the more massive you are, the higher the terminal velocity. For a human on Earth, the terminal velocity is around 195 kilometers per hour, and that is rather fast. You would reach that speed in about twelve seconds after falling about 450 meters. In general, hitting the ground at 195 kilometers per hour will kill you instantly, but you do not need to reach terminal velocity to do serious damage. Jumping from the top of a tree is bad enough.

There have been some extraordinary stories of survival. Juliane Koepcke, a German-Peruvian biologist, survived a fall from about three kilometers after the plane she was in was struck by lightning and exploded over the Peruvian rain forest in December 1971 during a severe thunderstorm. She fell to the ground still strapped into her seat and merely suffered a gash to her right arm and a broken collarbone. Other anecdotes tell of World War II pilots surviving high falls from aircraft into snowfields. However, these are rare anomalies.

Compare this challenge of surviving falls with that of an ant, whose mass gives it a terminal velocity of about six kilometers per hour, about thirty times less than for a human. Most ants can survive a fall at this speed with little damage.

Now let us increase the gravity, another factor in this equation, on an exoplanet to ten times that of Earth. The terminal velocity correspondingly increases. For a large animal, this will merely add to its woes in jumping off obstacles or falling from flight, but for a small insect, the terminal velocity, although now higher, may still

be sufficiently low to prevent a great deal of damage. An ant's terminal velocity, assuming the same density of atmosphere, increases to
about twenty kilometers per hour, still lower than that of a mouse
on our world. These ants can reach terminal velocity, land, and walk
off little the worse for their experience. In this return to the ants, we
have an illustration of how gravity is a less significant force for small
things. A distant super-Earth with a high gravity would have little
influence on its smallest forms of life.

In these examples, we see no compelling reasons to think that
rerunning the tape of evolution on another planet would give rise
to entirely new and unrecognizable forms of life. Instead, we would
probably witness the immutable laws of physics merely shaping life
in ways predictable in the context of the universal laws in which life
operates. These laws would probably give rise to forms different in
their minutia, but still narrowly channeled into a limited set of solutions, many of which would be familiar to us.

But let's have a little more fun. Sticking with gravity, we might
consider for a moment how this factor would affect flying creatures,
our alien starling and geese equivalents.

Whether you are watching a plane taking off from Edinburgh
Airport or the murmurations of starlings we have already met, these
flying contraptions must maintain the lift to stay airborne. We can
work out the force needed to stay aloft using the lift equation:

$$L = (C_L A \, \rho v^2)/2$$

where the lift (L), which is the upward force that keeps
the object in the sky, is the product of the surface area of
the wing (A) and the density of the air (ρ) through which the
object moves and the velocity (v) squared.

You will also notice another strange term in the equation. C_L
is the lift coefficient. This sort of thing is sometimes politely called
a fudge factor. It is not some sort of fundamental constant, but it
mops up all the complexities this equation cannot quite deal with:
the problem that lift is not merely related to the surface area of the

wing, but also to the wingspan, how the air interacts with different materials the wing is made of, and the angle of the wing (the angle of attack). C_L is something you must work out by doing experiments, and that is why aerodynamicists put model aircraft into wind tunnels—so they can determine what number they must use to get the right answer.

The equation tells us some simple things. The thicker the atmosphere, generally the more massive the creatures you can get off the ground. The lower the gravity, the less the creature is yanked down, and so again, we can get more-massive animals airborne.

There is no better place to illustrate the strange consequences of this simple physical principle than on Titan. This remarkable Saturnian moon is a mere 5,152 kilometers in diameter, 40 percent Earth's diameter, giving it a gravitational acceleration of 1.34 meters per second per second, just 13.7 percent of Earth's gravitational acceleration. Yet its thick atmosphere has a density of about 5.9 kilograms per cubic meter, compared with Earth's more meager 1.2 kilograms per cubic meter.

These numbers have not gone unnoticed to the imaginative. By reducing the weight of a person by over seven times but increasing the lift because of the higher density of the atmosphere, we suddenly arrive at the prospect of human flight.

Consider a typical human, say, with a mass of 70 kilograms. The person's weight on Titan is about 94 newtons. We must achieve this value or greater in lift if we are to fly. Using the lift equation, we might assume they are flying at a leisurely 5 meters per second (about 18 kilometers per hour). The lift coefficient we will take as 0.5 (a typical value). Using the density of Titan's atmosphere, we arrive at a required surface area of the wing of 2.5 meters squared. A wing this size will easily fit onto a wearable suit. Humans could jump off a cliff on Titan (in a spacesuit, of course) and soar in its skies like birds in slow, dignified, and elegant swoops.

Even on our own planet, with enough speed, people can fly using wing-suits. But on Titan, the lower gravity and thicker atmosphere mean you can glide slowly and gracefully without all that YouTube drama and death.

In these simple considerations, we have an example of how although the laws of physics are unyielding, small changes in the characteristics of distant worlds may open up evolutionary possibilities.
Sometimes, large planets with thin atmospheres may preclude atmospheric flight altogether, yielding biospheres where flight is limited to
short bursts into the sky, like flying fish or flying squirrels on Earth.
On other worlds, thick atmospheres on small planets could lead to
skies filled with flying creatures of immense size and variety.

Could these differences cause something drastic to happen,
something in evolution we were not expecting? Imagine that on a
distant exoplanet, smaller than Earth but with a thick atmosphere,
life had evolved. On that planet, large organisms, about the size of
humans, had come into existence, and not only that, they had ac-
quired intelligence. But there is something different about these creatures: they are of a birdlike lineage. They have wings.

The airborne capacities of these sentient fliers have had profound
consequences for their history. From the beginning of their recorded
time, they have been able to circumnavigate their planet and travel
immense distances. They do not commute to work on long roads,
but rather they fly. And so, the invention of automobiles was never
developed with any enthusiasm. Like all sentient beings, they have
some aggressive tendencies, but their ability to fly, to see other people and conurbations from far up in the sky, gave them a perspective
on other life and dulled their tendency for destruction. This view of
entire landmasses spurred an ecological and environmental awareness in their most primitive ages. The effect was that an effort started
from the nascent days of discovery of the scientific method to map
their entire planet and its environmental systems. This effort created
a planetary-scale sense of camaraderie in their species.

Their ability to fly meant that they quickly grasped the concepts
of artificial flight. Merely by observing their own wings during boring classroom lessons, early teenage geniuses figured out how the airfoil worked, leading to artificial flight. As flying hundreds of beings
across their planet for tourist or business purposes still had a use
alongside their natural capacities, they developed aircraft early.

Now, they generally led a peaceful existence because of their early environmental awareness. However, when war occurred, and sporadically it did, it was highly destructive. The beings' ability to drop objects from high above onto an enemy brought aerial warfare to their world when their society was young.

An important impetus brought on by their wings was spaceflight. With their natural tendency to see the world in three dimensions, their supernal view of society meant they quickly dreamed of flying beyond the atmosphere. As soon as they gathered the skills to mold metals and manipulate basic chemistry, they were already experimenting with rockets. A spacefaring capability rapidly followed their construction of aircraft.

I have taken you on this slightly bizarre excursion merely to illustrate that the claim that physics greatly constrains the forms of life does not mean we cannot generate an extraordinary variety of possible outcomes. I use the rather speculative example of intelligence to show that even though physical principles are the same everywhere, a small change in the physical conditions on a planet, in this case, a thick atmosphere, might well lead to biological outcomes and trajectories that have all sorts of indirect effects down the line, none more tangible to you and me than cultural implications. I have used the example of gravity to show that physical principles, given expression in equations, can be used, albeit rather simply in the examples I have chosen, to explore the potential effects on a hypothetical biota.

The discovery of exoplanets has shown us that the universe is full of a vast assortment of planetary possibilities. None may be exactly like Earth. Subtle differences in gravity, atmospheric density, landscapes, land-to-ocean ratios, and other factors will all influence the scope and content of any evolutionary experiments, if they exist. None would generate an exact replica of Earth's biosphere. The infinite diversity of small modifications in color and shapes will produce a fantastic and bewildering diversity of creatures. However, throughout these biospheres will run—at the small scale—the same architecture, the same construction from complex carbon molecules, the same repeating motifs in their major molecules, and the same compartmentalization. And at the large scale, they will run the same solutions to cope with gravity, to take to the air, or to swim in the sea. The equations of life permit variety and contingency within the underlying tapestry of striking conformity on Earth, and if life exists elsewhere, there as well.

For now, there is no immediate prospect of studying life forms in distant exoplanet biospheres directly, expanding the sample size of evolutionary experiments we can analyze beyond one. Only life in our own Solar System can provide us with that prospect in the near term. Regardless of the extent of our success or lack of it in finding an independent evolutionary experiment to explore, the discovery of exoplanets and the characterization of the environments they host will increasingly provide us with empirical knowledge of the diversity of physical conditions on rocky worlds. This new panorama will enrich and fuel our thoughts about how the characteristics of our own evolutionary experiment might have been different if it had been run on these distant worlds. From this vantage point, we may better understand how the physical conditions of Earth have fashioned the forms of life that have evolved here. With this new vista, we can deepen our insights of what might be universal about life.

CHAPTER 12

THE LAWS OF LIFE:
EVOLUTION AND PHYSICS UNIFIED

THAT THE LAWS OF physics and life are the same should surprise no one. The dissipation of energy into equilibrium in the universe is an inexorable process that allows for local complexity, of which life is a part. However, ultimately even the material within these oases of molecular diversity and the planet on which they reside must themselves disperse into the cold abyss. At the grandest scale, life is a flicker of light, eventually to be extinguished by one of the universe's most insistent physical laws, the second law of thermodynamics.

It is easy to appreciate our reluctance to allow physics to creep into the variety and color of life. Not only does physics bring to many people a terrifying reductionism, a cold, calculated view of the safari of living things that abound on Earth, but it also weakens a historical view that life differs from inanimate matter.

For many centuries, the idea that life is imbued with a vitalism, a force or substance that gives life its characteristic energy and unpredictability, was an essential way to build a barrier between the physical world and the domain of life. Without this barrier, living things and, dangerously therefore, people might well be relegated to a gray intersection with inanimate objects. It is this separation that has largely been responsible for the delay in understanding the evolutionary experiment in the context of physical principles. Throughout this book, we have explored how these principles variously restrict life at all levels of its hierarchy. The influence of physical principles on life should come as no shock. But if we briefly explore the history of the dichotomy between people's historical view of life and their understanding of physics, we can better understand why it has taken so long for life to be understood as a process embedded firmly in the same unassailable laws as those governing inanimate matter.

For centuries, the notion of spontaneous generation lay at the core of the difference between inanimate and animate matter. To move from the nonliving to the living, an elusive force must somehow transform material. In an age before peer review, science academies, and the general milieu of scientific discourse, no end of unusual and deeply strange recipes were on offer. Consider this one from Jan Baptista van Helmont, who in 1620 published a cookbook-like instruction for making mice:

> If you press a piece of underwear soiled with sweat together with some wheat in an open mouth jar, after about 21 days the odor changes and the ferment coming out of the underwear and penetrating through the husks of the wheat, changes the wheat into mice. But what is more remarkable is that mice of both sexes emerge, and these mice successfully reproduce with mice born naturally from parents. But what is even more remarkable is that the mice which came out of the wheat and underwear are not small mice, not even miniature adults or aborted mice, but adult mice emerge.

Many people would try to refute spontaneous generation. Experiments with animals were easy. In the seventeenth century, Francesco Redi, an Italian doctor, showed that by covering meat with a gauze, it would no longer carry out its famed spontaneous transformation into maggots, yet the gauze should have allowed a vital force through. The next step, to demonstrate that flies were necessary to make maggots, came rather readily.

However, microbes made life hard for the newly emerging scientific consensus. Just over sixty years after van Leeuwenhoek's discovery of microbes, John Turberville Needham, a well-respected scientist of the eighteenth century, reported his mutton-gravy experiment. Transferring some of his gravy from his household fire into vials and then closing them with stoppers, he reported how the gravy, after it had been sealed from the outside world, later teemed with life. He surmised that spontaneous generation was proven. The organic matter of the gravy had been infused with a vegetative life force.

We now know that it was likely that his gravy became contaminated with microbes. Lazzaro Spallanzani, better known for his pioneering studies on the regeneration of organs in frogs, repeated Needham's broth experiments, but he was more careful. After placing wetted seeds into vials, he sealed them and then heated them to kill anything that might still be alive. In vials heated for short periods, large animalcules died quickly. We now think these organisms must have been amoebae. He noticed that smaller animalcules, which were probably bacteria, could tolerate heat for many minutes until even they no longer moved. He then showed that if he heated his vials for long enough, they could be turned into "an absolute desert." He had demonstrated the concept of sterilization.

Now you might think this pioneering set of experiments would finally put the spontaneous-generation lobby out of business, but not so. It was easy for proponents to argue that the failure of life to be generated was because the material in the vials lacked access to air. By sealing his vials, Spallanzani had denied the organic matter the very gases needed for the life force.

With this historical backdrop, the now legendary Frenchman Louis Pasteur entered the stage. He had a mind of wonderful clarity for planning experiments. He would pioneer the process of pasteurization, in which the rapid and short-lived heating of milk could kill off microbes without changing the taste and so help preserve it. His response to this centuries-old question was an experiment of ingenious simplicity. He made swan-neck flasks, whose tapered ends elegantly curved to the side in a meandering S shape, denying the microbes the chance to drop directly into his broths, yet providing the liquid with oxygen. By the end of the nineteenth century, as well as destroying the last credibility of spontaneous generation, he had also demonstrated how to make a liquid sterile and keep it that way.

Embodied within this history is a deep and passionate conviction that there is something different about life. Even after the remnants of spontaneous generation have long since been dispersed in the storm of scientific progress, there remains an enduring feeling that biology should never merely be reduced to something driven in shape and form by physical processes. To accept that humans emerged from apes was difficult enough to bear after the exceptionalism of our Solar System was dismissed by the Copernican revolution, but at least apes stand with us, in biology, not of the inanimate realm. Darwin's evolutionary ideas could be explained by the hand of the Creator using evolution as a tool to achieve an ultimate end. Evolution was the mechanism behind Creation, but it was still special.

The successful union of biology and physics has been hampered not only by a long history that seeks a comfortable, special home for life, but also by a difference in culture and approach within the subjects themselves. I remember when I joined a physics department as a person who had spent most of his professional life among life scientists. Something startling did stand out to me. Sit down with a biologist and talk about a collaboration, and the first thing they do is start at the top, maybe talking about a microbe. Then, depending on the question you ask, they will work down into lower hierarchies in search of the answer. This habit probably stems from the vast complexity of ecosystems and living things in general. Trying to explain a mole's biology in terms of its constituent subatomic particles would seem a ridiculously futile task. Better to start with the mole and then try to answer questions at the next level down, its basic structure and what appendages it has. A biologist's proclivities to work from the top down are understandable.

Grab a cup of tea and start a discussion with a physicist, and you will often see the polar opposite. Their instinct is to start at the bottom and try to construct a simple model, perhaps manifested in an equation, of the process of interest. The proverbial spherical cow comes into view. This habit too is understandable. A field whose history has been to investigate the physical basis of the world around us and to express those features in mathematical relationships seeks to build, like a house, an edifice of knowledge from fundamental principles upward.

Neither approach is wrong. In fact, both approaches seem eminently suited to their subject matter. Physicists seek to acquire definiteness by working up to levels of a hierarchy where the behavior of lumps of matter can be encapsulated within equations. Biologists seek certitude in the face of extraordinary diversity and complexity in the biosphere by working reductively downward into things that can be more easily teased apart. But in these two approaches, there seems to be evidence of two very different sets of material. The two fields seem quite antipodal. To better understand how these apparently diametrical cultures have emerged, we should briefly explore the material that both sets of scientists grapple with.

If you rummage around at the bottom of the hierarchies of matter, its physics can become indiscernible. As Heisenberg revealed, small things are slippery customers. Measure the position of a subatomic particle, and you do not know what momentum it has. Measure its momentum, and its position is obscure. This fundamental property of particles, the Heisenberg uncertainty principle, is a behavior of the quantum world, the world of the miniature. Particles of matter, particularly subatomic ones, are not discrete entities that exist in one particular location, like a table or a chair. Rather, at this infinitesimal scale, they have wavelike properties like light, and these properties give their position in any place as a probability, not an exact value. The quantum world harbors other strange properties that seem quite alien and unfamiliar to our usual experiences.

But at the large scale, the variations in lots of different particles, say, the atoms of a gas, average out and we really can know something about objects, the scale of the universe familiar to you and me. Those uncertainties vanish in the sheer number of particles at this larger view. We can write down simple equations, like the ideal gas law, which allows us to predict the relationship between the pressure, volume, and temperature of a gas:

$$PV = nRT$$

where the pressure (P) and volume (V) of a gas is related to the number of moles of the gas (n), its temperature (T), and the universal gas constant (R).

The equation gives us certainty, whatever the uncertainties that swirl around the individual atoms of that gas. The idea is that physics has indiscernibility at the smallest scale, but it becomes less so when we look at things at the higher levels of a hierarchy. This is one reason why physicists (except those whose job is to understand the quantum world) often seek to describe phenomena by moving up the hierarchy to a vista where equations can capture the general behavior of matter.

Now compare this approach to how biologists look at living things. At the small scale, the structure of living things seems much simpler than the zoo of life we observe at the large scale all around us. In the machinery of the cell, there is predictability: molecules fold up according to principles of thermodynamics, binding occurs between base pairs in the genetic code in apparently predictable ways driven by simple chemical considerations, and energetic pathways that are ancient can be explained by thermodynamics. It all seems so far removed, so much more containable, than the diversity and endless forms we see in the world of whole organisms. At this larger scale, biology seems unpredictable. A vast collection of evolutionary creations makes up the biosphere and bedazzles the observer with its apparently limitless variety, a consequence of the creations' different histories and contingent details.

With this in mind, is it not surprising that biologists seek to escape this deluge of information by going after more tractable principles that operate at a level or two down in the structure of life? We might reasonably conclude that biology is predictable at the small scale, but it becomes whimsical and unpredictable at the scale of whole creatures. In contrast, physics is indiscernible at the small scale, but more predictable when Heisenberg's uncertainties and the oddity of quantum behavior seem to take back stage at the macroscopic scale. Biology is irretrievably the opposite of physics.

Although there is merit in this view, for me there is an equally compelling alternative, a perspective that emphasizes the unity between both fields and the similarity in the material that is their subject of study.

At the small scale, just like physics, biology does in fact have its uncertainties. Although the genetic code and the proteins that usher forth from its translation appear to be more predictable, less contingent than once was thought, biology is in one important respect less discernible at the small scale. The genetic code, faithfully reproducing information, generation after generation, is subject to change, to mutations.

Ionizing radiation, including natural background radiation, is one source of these changes. UV radiation from the Sun can do its bit to damage DNA. The energy it imparts to the code can cause adenine bases to bind together to form twins, known as pyrimidine dimers. When the genetic replication machinery meets these stuck-together bases, it misreads and introduces an error.

Chemicals too can damage DNA, including the mutation-causing carcinogens in cigarettes. Even more surprising is that no radiation or malign chemistry is needed to make mutations in DNA. They can occur quite spontaneously. Some bases (adenine and guanine) can fall apart, so to speak, dropping out of the double helix. Come replication time, those holes in the code cause an error in the new strand of DNA.

All of this shows us that unsurprisingly, DNA, like all machines, is not perfect. As it is exposed to the vagaries of the environment, natural chemical processes, and its imperfect replication, errors in its code are introduced. As we cannot predict exactly where in the code these mutations will occur (although we can define the susceptibility of certain molecules to damage of different kinds), there is an inherent unpredictability at the atomic and molecular level in how the code will alter over time. Most of these uncertainties are not quite the same as the uncertainties I have been speaking about in the quantum world. They are rather capricious changes caused by the imperfections of a genetic machine operating in a natural environment full of unexpected interferences from chemicals, radiation, or its own inherent weaknesses.

However, uncertainties familiar to physicists and biologists might well sometimes be one and the same, bringing the two fields into complete congruence. It was Per-Olov Löwdin, a Swedish scientist, who proposed that some mutations in DNA might be caused by quantum effects.

The binding of the base pairs down the center of the DNA double helix depends on bonding between the hydrogen (a proton) on one base and the oxygen or nitrogen atoms on its neighbor on the opposing DNA strand. These hydrogen bonds keep the two opposing strands of the DNA double helix together. They are the same bonds that come apart when the cell unzips the DNA down the middle to replicate the molecule during cell division or to translate it into protein.

Now sometimes that proton involved in a hydrogen bond, for example, on an adenine, can swap partners and leap across to the thymine partner on the other strand of the DNA. This proton swapping is rather problematic because now, with a proton relocated to a new molecule, the DNA replication machinery can become confused. When it meets the modified adenine, rather than attaching a thymine to it in the newly synthesized DNA strand as it should do, it may instead erroneously bind a cytosine. Now a mutation has been introduced; the code has been corrupted.

What was intriguing about Löwdin's suggestion was the mechanism for how this might occur. Getting a proton to jump across the middle of the DNA double helix from one base to another is no easy matter. Like driving your car over a hill on your way to the supermarket, you need to put in some energy. You need to step on the gas to get your car over to the other side. So too in chemistry. The proton must jump over a metaphorical hill that represents the energy needed to make that chemical change occur. It needs some energy to do this. But imagine that some generous neighbor has made a convenient tunnel through your local hill. Now, instead of using fuel to get over the hill, you drive effortlessly through the tunnel to the other side.

In the odd and strange quantum world in which subatomic par-ticles reside, precisely this quantum shortcut can occur. Rather than have to jump an energy-intensive hill, our peripatetic proton can quantum-tunnel from one base to another, more effortlessly making the transition. Löwdin theorized that quantum effects might lie at the heart of some mutations. His notion has been subjected to many efforts to model it and it remains something of a curiosity. Even if quantum tunneling does occur, is it common or even important? Nevertheless, I raise the question because it shows that at the small scale, some of the uncertainties in biology and physics may come from a common quantum source. Indeed, the entire field of quantum biology is built on investigating the physical uncertainties that unify the behavior of the smallest components of all matter, whether the matter is a gas or a goose.

Just like the case in physics, as we zoom out from all those un-predictable mutations, the picture differs and we can find common themes at the higher levels of organization too. At the large scale, all those capricious mutations, like the myriad of positions and mo-tions of atoms in a container of gas, average out. We end up with an organism that conforms to large-scale laws regardless of what may have happened at the atomic or molecular level. A mole is subject to $P = F/A$, and its cylindrical, pointy-faced shape, designed for opti-mizing the force it can apply to burrow and scrape its way through the soil, leads to convergent evolution. No matter how many unpre-dictable mutations have occurred in whatever bases in the genetic code of different moles, unless the changes are lethal, they do not obviate the requirement for those moles at the large scale to conform predictably to the laws that dominate their subterranean lifestyle.

In both evolutionary biology and physics, the material with which they work has uncertainty at the atomic scale, which smooths out at the larger scale to make systems of matter that take on forms grossly predictable. The two fields are thus unified.

I must, however, concede that there is one distinction that does exist between biology and physics, between living things and inani-mate matter. One crucial difference between life and nonlife is how all those uncertainties at the atomic and molecular scale change things at the grander scale. As Jacques Monod so eloquently pointed out, at the small scale, variations in most matter are often the source of its eventual deterioration and destruction. A small defect in a crystal might cause it to crumble away. An atomic displacement in a metal might be the source of a weakness, ultimately the birthplace of its structural failure, as bridge builders are only too aware. Crystals can develop a defect, but they have a problem. There is generally no way to reliably continue that defect into later generations of crystals so that we can find out whether, under certain circumstances, it might actually provide a way for the structure of that crystal to persist for longer than another crystal that has no defect. Because the inanimate object does not reproduce, there is no easy way for us to scatter small bits of that crystal into far-flung environments that have different physical and chemical conditions to see in which of those myriad places the defect provides an advantage that would then be passed on to later crystals while many of the rest of them just crumble into dust.

However, in life, those small mutations at the molecular scale in the DNA are the source of its variety. From these alterations in the code, a chance transformation in a base pair of DNA caused by the mischievous wiz of an errant high-energy particle, variant codes are produced. The code within life appears to give it a persistent purpose—"Life will find a way." However, this sense of purpose is an illusion. Life persists because differences in the code produce, in every successive generation, individuals, each with its own many idiosyncrasies. With so many variants, it is likely that some of these will survive and succeed in an environment, replicating with aplomb and filling an environment as far as is possible using the available resources, while elsewhere its kith and kin might perish. This process of selection gives the illusion of a sort of purpose or tenacity of life, a determination that somehow animates the living form. This character of life, the behavior that emerges from its code, does give it a special feature, but not one that categorically separates it from physics. This feature just makes life a particular embodiment of physical process, a coded process.

People are wont to imbue this chasm between life and nonlife with some sort of mystic unfathomability. Within this departure in the behavior of life from the rest of the cosmos, perhaps some see an opportunity to seize again that age-old desire for vitalism. Some people might hope to escape the nasty conclusion that life is just an interesting branch of organic chemistry, a particular, albeit fascinating, set of physical principles given expression in a special collection of molecules. Sadly, for those who dream of segregation, the difference is not that astounding.

One of the most fascinating things we can do in biology and physics—indeed, exactly what we have done in this book—is to take a journey from the sociobiology of ant nests to the atoms that make up life. Every level of the hierarchy, it is safe to say, is occupied by a different group of people (peruse the citations at the back of this book, and you can see this for yourself). Yet if you survey this literature, a common theme runs through it. Independently, many groups keep returning to the same idea. Look at life's choice of amino acids, and you find it rooted in the physical properties of those molecules. Observe the folding of proteins, and you find an almost infinite possible number of amino acid chains collapsed into just a few forms. Examine the structure of life, and cells are universal. Survey the forms of animals and plants, and simple relationships confine their shapes like rivets. Marvel at the arrangement of birds, ants, and fish, and within their bewildering hordes, simple rules are at work. From the most diminutive parts of life to whole populations, physical principles have been shown to hold life captive, to corral it into a small set of possibilities. For the last several decades, a triumph of biology has been to reduce vast complexity and ostensibly unfathomable diversity to apparent simplicity, aided by a newly formed convergence between biologists and physicists.

These groups of scientists, like so many laboratories working on different floors of a high-rise building, each occupying and researching a different level of the structure of life, seem to conclude the same thing. Life is confined by rules that are surprisingly, maybe shockingly, narrow. As we improve our grasp of how these rules govern biological systems, we may even predict, if not in fine detail at least in general, the pillars and girders of life. Synthetic biologists have entered the twilight zone of predicting the result of new forms of life with their own designer genetic codes, even creating these new forms too.

If biology and physics are thus united, is there any room for contingency in all this, for the quirks and fancies of history, the unpredictable leaps and bounds of evolution from previous forms that lurch into new territories of biology? Although I have suggested that these opportunities are rather limited, others have notably taken a very different view.

Stephen Jay Gould famously took sides with contingency, at least at the scale of whole organisms. For him, chance was everything in evolution. He recognized the underlying laws of physics, but believed adamantly that although these principles operate in the background, everything interesting in evolution was ultimately the result of contingency, the success of particular animal body plans, the rise of mammals, or the emergence of intelligence. These might have been thwarted had the dice rolled differently. His view lay in his experience with the Burgess Shale, a 508-million-year-old fossil deposit buried in the slopes of the Canadian Rocky Mountains. He elaborated on this perspective in his book *Wonderful Life*. Within these sheeted rocks are the imprints of some of the earliest animals, one of the best-preserved early experiments in multicellular, complex life. Repeated in similar units around the globe, the fossils bear witness to life's prolific steps beyond their three billion-year microbial past. Some of the strange arrangements of the segments, legs, tentacles, and other appendages in these animals seemed to suggest a biological landscape of contingency. From these unfamiliar forms, so Gould argued, the survival of our own lineage was a chance toss of the coin, a fluke.

I do not doubt that for any scientist who spends time analyzing the fine anatomical detail of these fossils—a bifurcated tentacle here, a segment strangely placed there, a leg doing something odd over there—their variety elicits sheer surprise, an example of contingency at work. However, as an outsider who has not studied the intricacies of invertebrates and become beguiled by them, but who did spend four days in a library poring over the reconstructions of the Burgess Shale monsters, I was struck not by the immense potentiality, but by their extraordinary uniformity. At that moment in history, there was the potential for an unlimited experiment; an open ocean (literally) existed for these newfangled animals to explore and exploit. Yet within their forms, they share a tedious similarity. Most are bilaterally symmetrical like you and me. Most have a mouth on their front and an anus on their rear. These animals have their detail and their bizarre contorted shapes, yet they also share page after page of gross similarity. They seem to be testament to an experiment so desperately trying to be unfettered, yet ultimately so unimaginative when confronted by the laws of hydrodynamics, diffusion, and a few other requirements besides. Contingency is there for sure, but it is none too stunning. The similarity in the products of this unique moment of unmuzzled opportunity was much more striking than any daring expeditions in biological form. Indeed, since Gould's paean to contingency, scientists have come to understand that many (if not all) of the Burgess Shale animals are related to modern groups of animals.

We must be careful to delineate two things. There is the contingency in the details of living things: the number of membranes that envelop a cell, the pattern of a moth's wing, the curves on the jaw of a dinosaur. In some refinements, if selection pressures do not vigorously impose themselves on the distinction between success and failure at reaching reproductive age, then contingency may have its way, a mixture of chance and some history. This very fact leads to the sheer enormity of variation in life on Earth.

For those who like the fine details, the variety and color of life, then I grant that contingency is everything. The historical nuances and developmental constraints of prior form may well in many features disallow the possibility of predicting the exact outcomes of evolution. However, on a deeper biological level, these are trivialities against the underlying physics that circumscribes life: the cell membrane, the aerodynamic forces that converge the wings of animals, the structure of jaws made for crushing food.

There is another realm in which contingency may play a role: in the major evolutionary transitions in life, the monumental changes that made a categorical difference to the capacities of life on Earth. Those transitions are not mere detail. Rather, they have shaped the edifice of life.

Some transitions seem to bear the mark of inevitability. The emergence of cellularity was required for biochemistry to be enclosed and then released within these comfortable chambers into the wider world. Without this step, life would never have emerged beyond a few localized self-replicating molecules in rocks or maybe a hydrothermal chimney.

Even in transitions that we could demonstrate are inevitable, their timing may be uncertain. Since an asteroid delivered that fateful blow to the dynasty of the dinosaurs 66 million years ago, the mammals have mutated from shrews to radiotelescope-building apes. Yet after about 165 million years of animal dominion over land, sea, and air, the dinosaurs had remained stuck in a reptilian state of intellectual dormancy. Not a space program in sight. At the very least, even if these animals would eventually have built dinosaur space agencies, the timing of some of the major events of evolution, perhaps intelligence, may well contain contingency, the chance imposition of the right selection pressure to drive forward cognition.

In the animals of the Burgess Shale and their fossil predecessors, Gould saw a specific moment of such transition-scale contingency at work. For him, the new forms of animals that so puzzled the first paleontologists who examined them postdate the potential for vastly different trajectories. Travel a little further back in time, and you arrive in the Ediacaran period, named for a picturesque location in Southern Australia. Here in the Ediacara Hills, the remains of the earliest animals known are preserved. All of them were soft-bodied, and most were flat, frond-like, quilted and pancake creatures. They precede the great "Cambrian explosion" that gave rise to the myriad of forms preserved in the Burgess Shale.

Why the flat animals? Animals, like cells, rely on a high surface area to take in nutrients and food and to exchange gases. In our own lineage, this feat is accomplished with internal organs. Lungs and intestines increase the effective surface area of each individual animal. Your lungs, with their networks of intricate tiny tubes, cover an area of about 75 square meters. Your gut, with all its coils and protrusions, absorbs food over a capacious 250 square meters, about the area of a tennis court. Yet in the Ediacaran, confronted by the same laws of physics, the need for food and gases to diffuse in and out, the solution was different. Instead, the creatures took on flattened forms so that no part of the animal was far from its surface. Gould asserted that if these Ediacaran faunas had inherited the Earth and outcompeted the solution of growing larger and using internal organs to bring life's substance into the body, the empire of the animals revealed in the gray slabs of the Burgess Shale might have been little more than a fandango of flattened forms. Contingency, the chance leap that evolutionary processes might have taken toward a different body plan when they reached that fork in the road, might have led to a very different world.

Can we say that the continued domination of the Ediacaran solution would have spelled doom for any further development of animals? Surely, one of the many lineages to have emerged from this experiment would sooner or later have developed an invagination within its body more capable of grabbing food and gathering oxygen? From that increase in surface area, such a form would have been at an advantage, perhaps allowing for more-complex and perhaps more-competitive animals?

Speculations are entertaining, but we cannot rerun this experiment. Whether there are contingencies, historical accidents of such import they could consign an entire biosphere to an ignominious eternity of pancakes, is an open question.

The science of evolutionary developmental biology and its revelations of both the hierarchical modularity of life and the way in which development can produce quite radical changes suggest that transitions of great magnitude, at least in whole organisms, can be made. The whale's pelvic bone, a vestige of its belligerent decision to write off the experiment of land-living and return to the oceans, is also the evidence of the extraordinary ability of life to dash from ocean to land and back again, flipping and changing to take advantage of, and adapt to, the equations of life. Perhaps the pancakes would eventually have become bloated and sprung forth legs.

Are there other contingencies between the first self-replicating molecule and a spacecraft-building civilization with the power to halt the complexity of the evolutionary experiment? Like the whale's indecision, the genetic code and metabolic pathways are apparently mutable. A growing compilation of evidence suggests that the inherent flexibility of even life's core processes allows it to escape frozen accidents and explore new possibilities, optimizing and improving under selection, even with the odd vestigial part remaining here and there. But what about the emergence of multicellularity or complex life?

Even if we can demonstrate the existence of such contingent events in the story of life, one or two major transitions that may well have hinged between two versions of the biosphere, possibly one world less complex than another, these to me are interesting, even startling moments. But they are mere solos, moments of singular diversion in a symphony of music, the music of physical laws. Their significance lies in that these contingent moments could, if they exist, determine whether intelligence is rare in the universe by deciding between two evolutionary paths, one of which leads to complex multicellular life and eventually intelligence, the other not. Such moments of chance that decide how quickly life reaches beyond its microbial beginnings, or whether it stretches beyond these limits at all, may have importance for the distribution of different levels of complexity of biospheres and their attendant life forms through the universe, if life exists elsewhere. However, these alternative worlds really only interest an intelligent onlooker who values intelligence, a species that finds these differences significant.

We could equally look on biospheres altered in their complexity by contingent events, maybe even in their possession of intelligence, as similar to the great variety of butterfly wings. The myriad possibilities evident in life on Earth and maybe elsewhere represent grand experiments in evolution, the phenomena of life. What is extraordinary about this branch of collected matter, the ideas we have surveyed in this book, is not what is contingent in it. Instead we marvel at how such a diversity and efflorescence of capacities, in a tiny bubble of physical and chemical conditions to be found in the known universe, can be so restricted.

The detail of life is titillating and a joy to gaze upon, but understanding the narrowness of the channels that ultimately hem life in is precisely what circumscribes a myriad of questions about life. Is carbon-based chemistry and the use of water as a solvent the only choice for a living thing? What makes metabolic pathways and the genetic code the way they are, and could they have been different? Why do cells look like they do? How does life adapt to new environments, and what limits are there? Why do animals have some solutions and not others, such as legs instead of wheels? How do physical limits shape the perimeter of the extremes of the biosphere itself? And beyond Earth, would other life, if it exists, look like terrestrial life? Endlessly onward.

Implicit in these lines of inquiry is whether contingency has a role to play. Although we cannot easily repeat evolution and observe contingency at work, we can study the factors that shape life using scientific observation and experimentation. Today, we can even modify the genetic code and explore its alternatives. Maybe one day, in exploring distant worlds, we will have another evolutionary experiment entirely to derive stronger ideas on the universality of biology. Life's predictabilities—its common features, its limits and boundaries—are what should enchant us.

Research that attempts to better establish how much physics constrains life and where chance may play a part, not merely in particular parts of life, but more synthetically across its whole hierarchy, has enormous potential to tighten the fusion between biology and physics. We can examine the physical principles that operate as information flows upward in the hierarchy from the genetic code to whole organisms. We can deepen our ability to describe evolution by more thoroughly investigating the principles that operate as environmental influences head downward, particularly in the form of selection pressures acting on organisms going about their lives and thence on the genetic codes that make it through to future generations.

Different levels of the biological hierarchy need not be linked to others. We need to define the physical principles that operate on life at a given scale independently of its properties at other scales to establish more exactly what is predictable about life. For instance, the fusiform, sleek body of a dolphin is sculpted by the action of hydrodynamics operating through natural selection on the whole organism. These selection pressures ultimately act through, but largely independently of, how that organism is assembled from the cellular level downward. Similarly, a fictitious alien mole-like creature shaped for effective burrowing would look similar to the tunneling creatures we are all familiar with, even if we were to imagine the animal assembled from a silicon-based chemistry. Many contingencies and inevitabilities at different scales are separate from those at lower or higher structures of organization. This understanding might greatly simplify our ability to predict or at least identify the physical principles at play in different parts of living things.

By expressing the physical principles within life in equations, we have the enthralling possibility of pushing ever more convincingly into the ability to predict structures and outcomes in the evolutionary process. As efforts intensify to expand this activity into an ever greater range of biological processes, so will the physical contours that define the nature of self-replicating, evolving systems of matter become more clearly resolved and more amenable to modeling and study.

This fascination with physical principles is not bland reductionism. In the finale to his seminal book on convergent evolution, Simon Conway-Morris laments the "dreary reductionism" of those who would incessantly try to simplify the complex structure of biology to genetic determinism.

Reductionism need not be dreary. There is beauty in physical simplicity. Is there not a stunning elegance and maybe even charm in physical equations manifested in the living form? An equation as unemotive and fundamental as $P = F/A$ reflected in the twitching nose and hurried digging of a mole, or the cold, hard rigor of buoyancy, $B = \rho Vg$, alive in the slender shapes and darting dances of a fish?

Around us is a biosphere of limitless and wonderful detail, but in its forms most simple. We see not a dreadful menagerie of three- and five-legged beasts, grotesquely fashioned in irregular shapes, a collection of creatures whose outlines bewilder and appall, a ghastly farrago of unbridled evolutionary contingency and experimentation. This is a biosphere of symmetry, of predictable scales and pleasing ratios, a pattern in form and construction that runs deep from the very core of biochemical architecture to families of ants and birds. It is the immutable and unbreakable marriage of physics and life.

ACKNOWLEDGMENTS

I HAVE BEEN VERY lucky over the years to move across a number of scientific disciplines without let or hindrance and to have observed some of the ferment occurring between them. I owe a debt of thanks to many institutions and people who during these years provided the impetus for exploring the ideas in this book. My undergraduate degree in biochemistry and molecular biology at the University of Bristol in the late 1980s exposed me to ideas on the underlying structure of life at a time when these sciences were getting into their stride, exploring their full reach with the new tools of molecular methods. A doctorate in molecular biophysics at the University of Oxford allowed me to see the newly emerging links between biology and physics, a synthesis of our understanding of living things and the principles of physics—two scientific directions that have tended to remain separate until only recently. The work going on at this interface provides a particularly thrilling vista on the basic principles that guide the assembly of life.

My transition into microbiology happened during my postdoctoral position at the NASA Ames Research Center in California, and it gave me a view of the microscopic world of life. I was especially fortunate to witness the birth of astrobiology when this science was taking on a new vigor in the form of the NASA Astrobiology Institute. As a great enthusiast of human space exploration and settlement, I could indulge my fascination with biology and space sciences at the same time. At Ames, I was influenced by many fine people who provided me with the opportunity to see the astronomical perspective on life, which one might say underpins the question of whether the structure of terrestrial biology can be described as universal, a question I pick up in these pages.

In moving to the British Antarctic Survey in Cambridge as a microbiologist to work among scientists whose focus was anything from penguins to seals, I developed a more ecological and evolutionary perspective on life. It was during afternoon walks among the nests of the South Polar skuas in the hills above Rothera Research Station on Adelaide Island, Antarctica, in what felt like an alien continent, when I began to wonder about the principles that shape the similarity between birds. The otherworldly backdrop of the "White Continent" provided the perfect environment to contemplate the extent to which all organisms are constrained into narrowly circumscribed forms, regardless of the place they call their home. I suppose this book is a development and a more in-depth reflection of some of these polar thoughts. Then, in moving to the Open University in Milton Keynes to work in a planetary sciences institute, I learned much about planetary-scale processes and their link with life. It is in this context that one is forced to think about how planetary conditions shape and channel the products of evolution. Now I am among physicists and astronomers at the University of Edinburgh, where a more reductionist view on life is a powerful undercurrent. All these places have been a pleasure to work in and a rich source of ideas. It has been immensely enjoyable to write a book that draws on research going on at the various levels of life's architecture that I have been privileged to observe firsthand.

In whatever way this book is received, I hope that minimally it might encourage a few additional evolutionary biologists and physicists to share a common interest in the remarkable products of evolution and inspire everyone else to look in awe at the beautiful simplicity of something that often seems so hopelessly complex.

I thank my research group at the time this book was written (Rosie Cane, Andy Dickinson, Hanna Landenmark, Claire Loudon, Tasha Nicholson, Sam Payler, Liam Perera, Petra Schwendner, Adam Stevens, and Jenn Wadsworth) for their forbearance. I also thank countless friends and colleagues who emailed suggestions or papers that in one way or another have influenced the thoughts expressed here. Thank you, Harriet Jones, Hanna Landenmark, Sydney Leach, and Rebecca Siddall, all of whom provided detailed comments on the manuscript. I am grateful to the University of Edinburgh, an excellent intellectual home for the last five years.

I am very grateful to my agent, Antony Topping, for his advice and guidance in developing this project. I thank T. J. Kelleher at Basic Books and Mike Harpley at Atlantic Books for their editorial suggestions and for guiding this book to its final publication.

NOTES

CHAPTER 1

2   *I once heard a distinguished:* This was an observation made by Martin Rees, Astronomer Royal, in a public lecture, but he also made a similar observation in print: "Even the smallest insect, with its intricate structure, is far more complex than either an atom or a star." Rees M. (2012) The limits of science. *New Statesman* **141 (May)**, 35.

2   *Other helium atoms:* Lequeux J. (2013) *Birth, Evolution and Death of Stars.* World Scientific, Paris.

3   *Like modern birds:* Witton MP, Martill DM, Loveridge RF. (2010) Clipping the wings of giant pterosaurs: Comments on wingspan estimations and diversity. *Acta Geoscientica Sinica* **31** Supp.1, 79–81.

4   *Scurrying among their short knobbly:* Edwards D, Feehan J. (1980) Records of *Cooksonia*-type sporangia from late Wenlock strata in Ireland. *Nature* **287**, 41–42; and Garwood RJ, Dunlop JA. (2010) Fossils explained: Trigonotarbids. *Geology Today* **26**, 34–37. Indeed, the type specimens of many early plants and invertebrates were found first in Scotland.

4   *Return just a few:* Pederpes: Clack JA. (2002) An early tetrapod from "Romer's Gap." *Nature* **418**, 72–76.

4   *Rather, in observing:* For folding of proteins, see Denton MJ, Marshall CJ, Legge M. (2002) The protein folds as Platonic forms: New support for the pre-Darwinian conception of evolution by natural law. *Journal of Theoretical Biology* **219**, 325–342. Working out how proteins fold is not a simple matter, a point raised with great clarity in Lesk AM. (2000) The unreasonable effectiveness of mathematics in molecular biology. *Mathematical Intelligencer* **22**, 28–37.

5   *Evolution is just:* This simple observation is compatible with the important role of natural selection in shaping life, but also with many factors that shape organisms that are not linked directly to primary selective effects. The multifarious ways in which organisms are evolutionarily shaped, explored for example by Gould and Lewontin, are entirely compatible with those same mechanisms being narrowly circumscribed and limited by physical principles. See Gould SJ, Lewontin RC. (1979) The spandrels of San Marco and the Panglossian paradigm: A critique of the adaptationist programme. *Proceedings of the Royal Society of London. Series B, Biological Sciences* **205**, 581–598. Some of those factors, particularly many "architectural" ones, are fundamentally physical constraints. For instance, Gould and Lewontin's spandrels are the physical consequence of joining two arches.

5   *The limited number:* For a remarkable discussion on the limitations and effectiveness of mathematics in describing physical processes, see Wigner E. (1960) The unreasonable effectiveness of mathematics in the natural sciences. *Communications in Pure and Applied Mathematics* **13**, 1–14. See Lesk (2000), above, for a more modern take on this classic essay.

6   *That the laws of physics:* A wonderful technical summary of some of the physical principles that are instantiated into life at the level of the whole organism is Vogel S. (1988) *Life's Devices: The Physical World of Animals and Plants.* Princeton University Press, Princeton, NJ. Steven Vogel also wrote a range of interesting papers examining fluid mechanics in life and other observations. The bibliography of his book, although dated, contains several excellent papers about physical measurements in organisms. A more popular exposition (although replete with detail and beautiful comparisons with human technology) is to be found in Vogel S. (1999) *Cats' Paws and Catapults: Mechanical Worlds of Nature and People.* Penguin Books, Ltd., London.

7   *At the molecular level:* Autumn K et al. (2002) Evidence for van der Waals adhesion in gecko setae. *Proceedings of the National Academy of Sciences* **99**, 12,252–12,256.

7   *The forces involved:* Alberts B et al. (2002) *Molecular Biology of the Cell* (4th ed.). Garland Science, New York.

7   *Put simply, when water is frozen:* Smith R. (2004) *Conquering Chemistry* (4th ed.) McGraw-Hill, Sydney.

7   *When clarifying how large:* Not all fish flex their bodies. Electric fish depend on keeping their bodies rigid so that they can generate stable

electric fields with which to sense the world. These fish have evolved a long, continuous fin along their body; the fin uses wavelike oscillations to drive the fish forward.

8   *And despite the inherent uncertainty:* Schrödinger's cat is a thought experiment in quantum mechanics elaborated by Erwin Schrödinger in 1935. The scenario involves a cat that may be simultaneously both alive and dead, made possible by a state known as a quantum superposition. It results from the cat's life being linked to a random subatomic event that may or may not occur. Werner Karl Heisenberg was a German theoretical physicist and one of the pioneers of quantum mechanics.

9   *The idea of organisms:* Or "fitness" landscapes. An elegant exposition of this concept, first developed by Sewall Wright, can be found in McGhee G. (2007) *The Geometry of Evolution.* Cambridge University Press, Cambridge.

11   *All these adaptations:* I am not dismissing the role of developmental constraints in evolution. See, for example, Smith JM. et al. (1985) Developmental constraints and evolution. *The Quarterly Review of Biology* **60**, 263–287; or Jacob F. (1977) Evolution and tinkering. *Science* **196**, 1161–1166. Indeed, very complex interactions can exist between physiology and evolution. See Laland KN et al. (2011) Cause and effect in biology revisited: Is Mayr's proximate-ultimate dichotomy still useful? *Science* **33**, 1512–1516. However, as will become apparent throughout this book, life seems to have more flexibility to overcome these prior historical quirks and "fixed accidents" than is typically assumed, whether they be in the genetic code or in macroscopic forms of creatures. That is not to say that we cannot find plentiful evidence of history in animals—such as the four legs of land-dwelling animals derived from pectoral and pelvic fins of fishes. This history may restrict the options open to life within what are the dominant physical principles that shape it.

11   *For simplicity's sake:* The reader might claim a certain degree of tautology here. Whether evolution is a characteristic of life rather raises the question of how we define life. We can indulge in fantastical ideas of life forms that adapt to their environment and read these adaptations back into their genetic code in a Lamarckian form of evolution. If such a system were powerful enough, the Linnaean system of classification we associate with the hierarchical nature of the phylogeny of life on Earth would not emerge. However, along with Dawkins R. (1992) Universal biology. *Nature* **360**, 25–26, I am going to start with

an assumption that evolution in a Darwinian sense is universal in natural things that replicate with a code. Indeed, here I will simply take it as a working assumption of my book that systems of matter that reproduce and exhibit Darwinian evolution are the things that concern my discussion. Even if the reader refutes this universality and can describe a reproducing system that adapts to its environment quite differently, most of the conclusions I draw in this book, particularly regarding the restricting effects of physical processes, are likely to hold. A real example might be very early cells on Earth when life first arose in which genetic information may have passed more fluidly between them just as horizontal gene transfer occurs in microbes today such as discussed in Goldenfeld N, Biancalani T, Jafarpour F. (2017) Universal biology and the statistical mechanics of early life. *Philosophical Transactions A* **375**, 20160341. Some people have argued that such a community of cells has non-Darwinian properties (in that genetic material is added into a primitive genome in a quasi-Lamarckian way). However we situate these ideas in a description of the evolutionary process (even the products of horizontal gene transfer are still subject to environmental selection), the processes are narrowly circumscribed by physics. Dawkins puts a compelling case that Darwinism is not merely part of the definition of life as we know it, but a universal characteristic of replicating things that have adaptive complexity: Dawkins R. (1983) Universal Darwinism. In *Evolution from Molecules to Men*, edited by DS Bendall, Cambridge University Press, Cambridge, 403–425. For Joyce quote, see Joyce GF. (1994) In *Origins of Life: The Central Concepts*, edited by DW Deamer and GR Fleischaker, Jones and Bartlett, Boston, xi–xii. Joyce points out informally on the internet that the definition was developed during panel meetings of NASA's Exobiology Discipline Working Group in the early 1990s.

11   *We could argue that the word* life: A problem explored by Cleland CE, Chyba CF. (2002) Defining "life." *Origins of Life and Evolution of Biospheres* **32**, 387–393.

12   *In his engaging 1944 book:* Schrödinger E. (1944) *What Is Life?* Cambridge University Press, Cambridge.

13   *Mathematical models:* Discussed by Baverstock K. (2013) Life as physics and chemistry: A system view of biology. *Progress in Biophysics and Molecular Biology* **111**, 108–115.

15   *As early as 1894:* Wells HG. (1894) Another basis for life. *Saturday Review*, 676.

15   *In 1986, Roy Gallant:* Gallant R. (1986) *Atlas of Our Universe.* National Geographic Society, Washington DC.

16   *What we see on Venus:* To emphasize the caveat I make in the main text, someone imaginative could argue that these planets just lacked an origin of life or that an origin of life is very rare. However, if life had originated on these planets, we would indeed see these very creatures. It is difficult, in the absence of any probabilities on the origin of life or a certainty about the conditions required for it, to argue against this position. However, as I will discuss in later chapters about the limits to life, there are more-fundamental limits to the possibility of life in hell-like worlds such as Venus, regardless of whether an origin of life could have (or even did) occur there.

17   *It is apposite:* Darwin C. (1859) *On the Origin of Species by Means of Natural Selection, or the Preservation of Favoured Races in the Struggle for Life.* John Murray, London.

CHAPTER 2

20   *Ant civilization:* Wilson EO. (1975) *Sociobiology: The New Synthesis.* Belknap Press, Cambridge, MA. The book Wilson wrote in collaboration with Hölldobler on ants was the first academic work to win the Pulitzer Prize: Hölldobler B, Wilson EO. (1998) *The Ants.* Springer, Berlin.

21   *Quickly, we have:* A more macabre demonstration of feedback processes in ant societies has been shown in how ants make piles of ant corpses: Theraulaz G et al. (2002) Spatial patterns in ant colonies. *Proceedings of the National Academy of Sciences* **99**, 9645–9649.

22   *Remarkably, no architect:* I have focused on the rules that drive certain collective behaviors in ants. Another question entirely is why ants live together in the first place and how eusociality (the tendency that some groups of animals have to be split into reproductive and nonreproductive groups, the latter merely tending for everyone else) could have arisen in the raw competitive world of evolution. This question can itself be reduced to plausible physical principles and mathematical modeling and is discussed in Nowak MA, Tarnita CE, Wilson EO. (2010) The evolution of eusociality. *Nature* **466**, 1057–1062.

22   *For example, for the ant species* Messor sancta: Quantified and discussed in Buhl J, Gautrais J, Deneubourg JL, Theraulaz G. (2004) Nest excavation in ants: Group size effects on the size and structure

of tunnelling networks. *Naturwissenschaften* **91**, 602–606; and Buhl J, Deneubourg JL, Grimal A, Theraulaz G. (2005) Self-organised digging activity in ant colonies. *Behavioral Ecology and Sociobiology* **58**, 9–17.

23   *Perhaps best known:* Willmer P. (2009) *Environmental Physiology of Animals.* Wiley-Blackwell, Chichester.

23   *The exact physical underpinnings:* However, some excellent papers explore the basis of these laws and themselves are usually based on physical models. One example is West GB, Brown JH, Enquist BJ. (1997) A general model of allometric scaling laws in biology. *Science* **276**, 122–126, which proposes that the basis of many of the physiological power laws in life are rooted in the need to transport materials through linear networks that then branch out to supply all parts of the organism. They use this supposition to develop a model that predicts a variety of structural features of living forms, from plants to insects and other animals.

24   *Many fixed relationships:* For a good account of these ideas and the past literature on allometric power laws and their physical basis, I very much recommend West GB. (2017) *Scale: The Universal Laws of Life and Death in Organisms, Cities and Companies.* Weidenfeld & Nicolson, London.

24   *Attempting to reduce:* The classic paper that proposed a model of simple particle motion that would make the transition from disordered to ordered behavior using some basic rules was Vicsek T et al. (1995) Novel type of phase transition in a system of self-driven particles. *Physical Review Letters* **75**, 1226–1229, and was applied to biological systems in Toner J, Tu Y. (1995) Long-range order in a two-dimensional dynamical model: How birds fly together. *Physical Review Letters* **75**, 4326–4329. The transitions that give rise to this sort of self-organized behavior were further elaborated on by Grégoire G, Chaté H. (2004) Onset of collective and cohesive motion. *Physical Review Letters* **92**, 025702. There are, of course, many other papers exploring the physics of self-organization applied to both nonliving and biological systems.

24   *This field strives:* Self-organization can be observed at many scales, not just in biology, but in all physical systems, including weather systems: Whitesides GM, Grzybowski B. (2002) Self-assembly at all scales. *Science* **295**, 2418–2421. For a nice short summary of how systems far from equilibrium are relevant to biology, see Ornes S. (2017) How nonequilibrium thermodynamics speaks to the mystery of life.

Proceedings of the National Academy of Sciences **114**, 423–424. His missive also contains some other relevant citations on nonequilibrium systems in biology.

26   *Like other aspects:* This formulation has been shown to predict behaviors in, for example, the Argentine ant (*Iridomyrmex humilis*): Deneubourg JL, Aron S, Goss S, Pasteels JM. (1990) The self-organizing exploratory pattern of the Argentine ant. *Journal of Insect Behaviour* **3**, 159–168.

28   *Like a miniature computer:* A discussion of the differences between ants and molecules, as well as principles of interactions between ants is Detrain C, Deneubourg JL. (2006) Self-organized structures in a superorganism: Do ants "behave" like molecules? *Physics of Life Reviews* **3**, 162–187.

29   *The reactions complicate:* Models can be made that take into account how memory, for example in bird flocks and schooling fish, affects subsequent group behavior. Random fluctuations that cause large-scale gross changes in animal groups can also be investigated. These attributes add complexity to models, but at their core, the models are still constructed on the basic principles of how the component organisms interact: Couzin ID et al. (2002) Collective memory and spatial sorting in animal groups. *Journal of Theoretical Biology* **218**, 1–11.

29   *Hampering efforts:* A paper that reviews this history as well as some of the theories on bird flocking is Bajec IL, Heppner FH. (2009) Organized flight in birds. *Animal Behaviour* **78**, 777–789.

30   *At the core:* A detailed paper looking at some of these assumptions is Chazella B. (2014) The convergence of bird flocking. *Journal of the ACM* **61**, article 21. Also see Barberis L, Peruani F. (2016) Large-scale patterns in a minimal cognitive flocking model: Incidental leaders, nematic patterns, and aggregates. *Physical Review Letters* **117**, 248001.

31   *Yet rules applied:* A model that examines how vertebrates can organize, find new food sources, or navigate to new places with only a few individuals in the group with access to the necessary information is Couzin ID, Krause J, Franks NR, Levin SA. (2005) Effective leadership and decision-making in animal groups on the move. *Nature* **433**, 513–516.

31   *The infant state:* In the case of bird flocking, a forceful paper that examines their collective behavior as a physical process (with a wonderful title that only a physicist can muster) is Cavagna A, Giardina I.

(2014) Bird flocks as condensed matter. *Annual Reviews of Condensed Matter Physics* **5**, 183–207.

32   *However, evidence:* This idea was first elaborated by Wynne-Edwards VC. (1962) *Animal Dispersion in Relation to Social Behaviour.* Oliver & Boyd, Edinburgh. One of the idea's problems is that it suggests a form of bird behavior directed to the good of the group (a theory that was at the forefront of Wynn-Edwards's writing), a form of self-censorship on breeding behavior. A bird that took part in the census but then cheated by having a few more offspring than other birds would quickly spread in the population, potentially vitiating the whole strategy. Furthermore, clutch (egg number) has not been shown to regulate in response to the numbers of birds in a murmuration, making the idea difficult to test empirically.

33   *Weimerskirch saw:* Weimerskirch H et al. (2001) Energy saving in flight formation. *Nature* **413**, 697–698.

34   *time, it was not for filming:* Portugal SJ et al. (2014) Upwash exploitation and downwash avoidance by flap phasing in ibis formation flight. *Nature* **505**, 399–402.

36   *Filaments are a little easier:* Schaller V et al. (2010) Polar patterns of driven filaments. *Nature* **467**, 73–77.

37   *Tim Sanchez:* Sanchez T et al. (2012) Spontaneous motion in hierarchically assembled active matter. *Nature* **491**, 431–435.

37   *About four times:* That's 0.000000025 meters.

37   *The rules and principles:* A comprehensive text that synthesizes information on self-organization in diverse organisms, including ants, bees, fish, and beetles is Camazine S et al. (2003) *Self-Organization in Biological Systems.* Princeton University Press, Princeton, NJ. The book also discusses the general reasons and principles behind self-organization, including its ability to enhance the formation of stable structures. The book contains an wide-ranging set of references to various works covering self-organization. A highly comprehensive study of self-organization is to be found in Kauffman S. (1993) *The Origins of Order: Self-Organization and Selection in Evolution.* Oxford University Press, Oxford, which is beautifully summarized in his popular science book: Kauffman S. (1996) *At Home in the Universe: The Search for Laws of Self-Organization and Complexity.* Oxford University Press, Oxford. And see the work by Ao: for example, Ao P. (2005). Laws of Darwinian evolutionary theory, *Physics of Life Reviews* **2**, 117–156.

38   *It is easy to think:* Despite our desire to consider ourselves separate from "mere" natural processes, human populations are amenable to modeling as well, such as this fascinating study of city size and shape shows: Bettencourt LMA. (2013) The origins of scaling in cities. *Science* **340**, 1438–1441.

38   *The self-organization of life:* Although I have focused on aspects of self-organization to illustrate physical principles at work, many other areas of physics and mathematics may be applied to understanding the operation of groups of organisms. One major contribution has been the biological and evolutionary application of game theory, which seeks to understand the evolutionary benefits of different choices taken by organisms—and for which there is a vast amount of literature. See Maynard Smith J, Price GR. (1973) The logic of animal conflict. *Nature* **246**, 15–18. A book looking at the application of game theory to biology is Reeve HK, Dugatkin LE. (1998) *Game Theory and Animal Behaviour.* Oxford University Press, Oxford. A thoroughgoing technical text that explores these evolutionary interactions and other aspects of the application of mathematical theory to evolution is Nowak MA. (2006) *Evolutionary Dynamics: Exploring the Equations of Life.* Belknap Press of Harvard University Press, Cambridge, MA. I discovered his book after the decision on the title of my book was long since committed. However, I feel no proprietary concern. The "equations of life," I think, is a natural phrase that succinctly captures the manifestation of life in physical principles given expression in mathematical relationships that can be written in equations. Moreover, *The Equations of Life* is the title of a novel by Simon Morden. Set in a postnuclear apocalypse, the book's plot involves a link between physics and evolutionary biology—a link that is perhaps best avoided.

CHAPTER 3

39   *In the winter of 2016:* And I'd like to thank the members of this group for their work, on which this chapter is based: Julius Schwartz, Hamish Olson, Danielle Hendley, Emma Stam, Rodger Watt, and Laura McLeod. They did a very fine job and wrote a splendid report.

40   *With so many degrees:* Cruse H, Durr V, Schmitz J. (2007) Insect walking is based on a decentralized architecture revealing a simple and robust controller. *Philosophical Transactions of the Royal Society A* **365**, 221–250.

40   *Wind speed:* The physics and mathematics of insect legs and locomotion is a fertile area of research, driven by an interest in creating legged robots that will more effectively navigate terrain. See, for example, Ritzmann RE, Quinn RD, Fischer MS. (2004) Convergent evolution and locomotion through complex terrain by insects, vertebrates and robots. *Arthropod Structure and Development* **33**, 361–379.

40   *The ladybug, like spiders:* Some insects have smooth pads.

40   *With it, we can predict:* The development of these models can be found in a number of papers, such as, Zhou Y, Robinson A, Steiner U, Federle W. (2014) Insect adhesion on rough surfaces: Analysis of adhesive contact of smooth and hairy pads on transparent microstructured substrates. *Journal of the Royal Society Interface* **11**, 20140499. The equation shown in this chapter can be found in Dirks JH. (2014) Physical principles of fluid-mediated insect attachment—shouldn't insects slip? *Beilstein Journal of Nanotechnology* **5**, 1160–1166.

41   *The first term is the surface tension:* The Laplace pressure is the pressure difference between the inside and the outside of a curved surface that forms a boundary between a gas and a liquid region. This pressure difference is caused by the surface tension of the interface between the two regions.

41   *To achieve this, the leg:* All biological structures, particularly appendages, are evolved to have factors of safety (the ratio of the stress that causes failure to the maximum stresses experienced). This is not to say that evolution has engineering foresight, but these factors are likely to minimize the probability of failure sufficiently not to significantly impinge on survival. For a comprehensive and interesting discussion, see Alexander RMN. (1981) Factors of safety in the structure of animals. *Science Progress* **67**, 109–130, which touches on the field of biomechanics, yet another field that brings together physics and biology, especially at the level of the whole organism, although Alexander also considers seeds and other biological structures.

41   *From the top:* Peisker H, Michels J, Gorb SN. (2013) Evidence for a material gradient in the adhesive tarsal setae of the ladybird beetle *Coccinella septempunctata*. *Nature Communications* **4**, 1661.

41   *equations:* Federle W. (2006) Why are so many adhesive pads hairy? *Journal of Experimental Biology* **209**, 2611–2621.

43   *Yet at the scale:* I do not exaggerate when I say that one of my favorite scientific papers, which explores this topic exactly, is Went FW. (1968) The size of man. *American Scientist* **56**, 400–413. Went draws

our attention to the different physics principles operating at the small and large scales and their biological implications, discussing the forces of gravity at the large scale and molecular forces that dominate at the small scale. Particularly entertaining is his thought experiment on the ant preparing to go to work. If you want his explanation on why the ant can't kiss his wife good-bye or have a sneaky cigarette on the way to work, you'll have to read the paper yourself. Another earlier paper in the same vein is Haldane JBS. (1926) On being the right size. *Harper's Magazine* **152**, 424–427. Here Haldane pays particular attention to insects and argues that the size of an organism mandates what sorts of systems it must have to exist. Implicitly, he is recognizing that physical size pulls into play physical principles that ultimately decide how a living thing is constructed, not mere contingency.

44   *However, we can unravel:* When I use the term *contingency* throughout this book, I mean an evolutionary development that was a quirk of history, a chance path that could have been very different. Stephen Jay Gould and other scientists who believe that contingency is an important driver in evolution theorize that if the tape of evolution were rerun, a completely different set of paths might be followed. Note some subtlety here. Contingency could refer to two similar or identical evolutionary experiments changed by chance mutations on their course, or it could refer to small, different historical conditions, such as at the start of an evolutionary experiment, radically changing the outcome of evolution. Usually in this book, I am referring generally to both possibilities.

44   *If the insect is distracted:* Jeffries DL et al. (2013) Characteristics and drivers of high-altitude ladybird flight: Insights from vertical-looking entomological radar. *PLoS One* **8**, e82278.

44   *Rapid advances:* I have deliberately not written equations for insect flight here since the equation of lift, which I use later, is too simple to capture the complexity of insect aerodynamics. To list one equation would also force me to list many more to even do the subject cursory justice. However, for details on the phenomenon, I refer the reader to the following papers, although there are many more: Dickinson MH, Lehmann F-O Sane SP. (1999) Wing rotation and the aerodynamic basis of insect flight. *Science* **284**, 1954–1960; Sane SP. (2003) The aerodynamics of insect flight. *Journal of Experimental Biology* **206**, 4191–4208; Lehmann F-O. (2004) The mechanisms of lift enhancement in insect flight. *Naturwissenschaften* **91**, 101–122; Lehmann F-O, Sane SP, Dickinson M. (2005) The aerodynamic effects of

wing–wing interaction in flapping insect wings. *Journal of Experimental Biology* **208**, 3075–3092.

45   *Its solution, chitin:* Mir VC et al. (2008) Direct compression properties of chitin and chitosan. *European Journal of Pharmaceutics and Biopharmaceutics* **69**, 964–968.

46   *The severity of collisions:* Henn H-W. (1998) Crash tests and the Head Injury Criterion. *Teaching Mathematics and Its Applications* **17**, 162–170.

46   *Well, yes, attracting a mate:* The formation of colors in the natural world, such as in the wings of butterflies, is an exquisitely developed area of physics covering photonics and other fields. Just one such paper is Kinoshita S, Yoshioka S, Miyazaki J. (2008) Physics of structural colors. *Reports on Progress in Physics* **71**, 076401.

46   *It was Alan Turing:* Turing AM. (1952) The chemical basis of morphogenesis. *Philosophical Transactions of the Royal Society Series B* **237**, 37–72.

47   *By varying the range:* A description of the use of the Turing model for explaining and predicting patterns has even been applied to ladybugs themselves: Liaw SS, Yang CC, Liu RT, Hong JT. (2001) Turing model for the patterns of lady beetles. *Physical Review E* **64**, 041909.

47   *But the essential idea:* Rudyard Kipling's writing preceded Turing's paper, but if Kipling had been born later, he might have collaborated with Turing in writing his *Just So* story "How the Leopard Got His Spots."

48   *Indeed, dark ladybugs:* Two papers investigating this effect are Brakefield PM, Willmer PG. (1985) The basis of thermal melanism in the ladybird *Adalia bipunctata*: Differences in reflectance and thermal properties between the morphs. *Heredity* **54**, 9–14; and De Jong PW, Gussekloo SWS, Brakefield PM. (1996) Differences in thermal balance, body temperature and activity between non-melanic and melanic two-spot ladybird beetles (*Adalia bipunctata*) under controlled conditions. *Journal of Experimental Biology* **199**, 2655–2666. For observations of the same effects in dark- and light-colored beetles in the Namib Desert, see also Edney EB. (1971) The body temperature of tenebrionid beetles in the Namib Desert of southern Africa. *Journal of Experimental Biology* **55**, 69–102.

49   *The little equation:* See De Jon PW et al. (1996), above.

49   *Keeping their temperature:* For a general paper on insect thermoregulation that also considers the role of shivering, see Heinrich B. Keeping

their temperature high enough to move around: (1974) Thermoregulation in endothermic insects. *Science* **185**, 747–756; and his later book Heinrich B. (1996) *The Thermal Warriors: Strategies of Insect Survival.* Harvard University Press, Cambridge, MA.

49 *The drop in freezing point:* The molality is the moles of a chemical divided by its mass.

51 *Instead, cylinders run:* An early paper discussing some of the principles for flying insects is Weis-Fogh T. (1967) Respiration and tracheal ventilation in locusts and other flying insects. *Journal of Experimental Biology* **47**, 561–587.

53 *Oxygen has:* See, for example Verbeck W, Bilton DT. (2011) Can oxygen set thermal limits in an insect and drive gigantism? *PLoS One* **6**, e22610. A very good paper that explores all the complex factors that may influence the role of oxygen in insect size is Harrison JF, Kaiser A, VandenBrooks JM. (2010) Atmospheric oxygen level and the evolution of insect body size. *Proceedings of the Royal Society B,* doi:10.1098/rspb.2010.0001.

53 *Larger insects:* For a thorough discussion of the problems confronting theoretical one-kilogram grasshoppers, see Greenlee KJ et al. (2009) Synchrotron imaging of the grasshopper tracheal system: Morphological and physiological components of tracheal hypermetry. *American Journal of Physiology. Regulatory, Integrative and Comparative Physiology* **297**, R1343–1350.

54 *The angular size of each lens:* Barlow HB. (1952) The size of ommatidia in apposition eyes. *Journal of Experimental Biology* **29**, 667–674.

55 *The reception of light:* The evolution of the protein receptors that gather light, the opsins, can occupy an entire tract unto themselves. They demonstrate convergence at the molecular level. So too the evolution of compound eyes and camera eyes. From the scale of the eye to its molecular components, the evolution of eyes is riven with convergence. Because the purpose of the apparatus is to capture electromagnetic radiation, physical principles have very strongly channeled convergence. See, for example, Shichida Y, Maysuyama T. (2009) Evolution of opsins and phototransduction. *Philosophical Transactions of the Royal Society* **364**, 2881–2895; Yishida M, Yura K, Ogura A. (2014) Cephalopod eye evolution was modulated by the acquisition of Pax-6 splicing variants. *Scientific Reports* **4**, 4256; and Halder G, Callaerts P, Gehring WJ. (1995) New perspectives on eye evolution. *Current Opinions in Genetics and Development* **5**, 602–609.

There is a plethora of other papers that investigate the details of eye evolution. All of them are a journey into a rich link between biology and physics. A substantive book on the equations of eyes would be entirely merited.

55   *We have not talked:* Weihmann T et al. (2015) Fast and powerful: Biomechanics and bite forces of the mandibles in the American Cockroach *Periplaneta americana. PLoS One* **10**, e0141226. The physics of the food that insects eat might itself influence the physics of insect mandibles required to eat it. For a discussion of the physics of grass, this splendidly titled paper is worth reading: Vincent JFV. (1981) The mechanical design of grass. *Journal of Materials Science* **17**, 856–860. Physical principles have been shown to be behind the evolution of the jaws of many organisms, such as the intriguing extinct giant otters: Tseng ZJ et al. (2017) Feeding capability in the extinct giant *Siamogale melilutra* and comparative mandibular biomechanics of living Lutrinae. *Scientific Reports* **7**, 15225.

55   *And the physics:* Already partly taken on in Gutierrez AP, Baumgaertner JU, Hagen KS. (1981) A conceptual model for growth, development, and reproduction in the ladybird beetle, *Hippodamia convergens* (Coleoptera: Coccinellidae). *Canadian Entomologist* **113**, 21–33.

55   *I suspect three:* Although the topic is not explored from the point of view of physics, an excellent text that does bring one face-to-face with the incredible complexity of insects and their extraordinary abilities is Chapman RF. (2012) *The Insects: Structure and Function.* Cambridge University Press, Cambridge.

56   *Natural selection:* Physical principles described in the forms of equations are ultimately mathematical, and so we would not be completely misled in describing life as merely the manifestation of mathematics (see du Sautoy M. [2016] *What We Cannot Know.* Fourth Estate, London, for a rather eloquent exposition of the idea that the universe is just a manifestation of mathematics). But here, I generally focus on physical principles to draw attention to the physical manifestations of life's mathematical relationships and equations and their impact on the form and function of organic matter with a code that evolves. At this point in the text, however, *mathematical* seems apposite to emphasize the interconnected mathematical relationships between the different terms of equations manifest in the living form.

57   *Such efforts would take us:* There are two forms of prediction. One is reductive prediction, for example, the ability to predict cell membrane

structure using the knowledge that living things are made from cells. In other words, prediction at lower levels of complexity according to knowledge at higher levels. The other form, predictive synthesis, is the ability to predict complex structures according to knowledge about lower levels of its hierarchy. The latter is generally much more difficult than the former since how simple things assemble into complex structures is less well understood. However, increasingly the capacities for prediction in both directions are being improved. An eloquent exploration of these forms of prediction is made by Wilson EO. (1998) *Consilience*. Abacus, London, 71–104.

58   *genes are involved:* The modularity of developmental processes and phenotypic characteristics may make this task more tractable than it might first seem. This modularity is reviewed in Müller GB. (2007) Evo-devo: Extending the evolutionary synthesis. *Nature Reviews Genetics* **8**, 943–949.

58   *Nevertheless, this evolutionary physics:* I refer the reader to a fascinating paper in which the authors attempt to capture adaptation using statistical physics, illustrating the sort of approaches that might aid a description of biological processes in the form of equations that can yield predictive power: Perunov N, Marsland R, England J. (2016) Statistical physics of adaptation. *Physical Review X* **6**, 021036. Regarding evolutionary changes in quantitative, physically circumscribed terms, what I describe here can be thought of as a type of optimality model (see, for example, Abrams P. [2001] Adaptationism, optimality models, and tests of adaptive scenarios. In *Adaptationism and Optimality*, edited by SH Orzack and E Sober. Cambridge University Press, Cambridge, 273–302). In the real world, of course, knowing which traits are important to an organism and what their optimized properties are is extraordinarily difficult, particularly for poorly studied organisms. Thus, these approaches have their limitations. Nevertheless, my suggestion might allow us to understand better certain physical trade-offs and create a more converged area of thought between physicists and biologists. This approach can work particularly well where empirical information is forthcoming, such as the heat balance in ladybirds and its link to wing-case thickness and metabolic rates in given environmental conditions. We may not know optimal conditions, but we can measure quantities in given organisms and use these measurements to define a parameter space of interacting equations that can be used to investigate how different physical constraints in the environment and physical quantities in organisms interact to influence its adaptive traits.

58   *Throughout this tantalizing foray:* I say "often" because some examples
      of convergence appear to be driven more by biological interactions
      than by the physical environment. Mimicry is an example found var-
      ious places, from butterfly wings to stick insects that look like twigs
      or leaves. Many of these examples are generally about the cosmetic
      appearance of organisms rather than their fundamental structure and
      mechanics. However, we might still think about them from a physical
      perspective. Although the selection that gives rise to these phenomena
      is driven by interactions between organisms, at a fundamental level
      a larva that evolves to look like a twig is an example of one lump of
      matter containing a code evolving to look like another lump of matter.
      This similarity means that the lump of twiglike matter called a larva
      is less likely to be destroyed and so its abundance correspondingly
      increases. This process is understood just as easily in simple physical
      terms. Once we think about organisms as lumps of matter with a code,
      the distinction between biology and physics becomes less distinct.

CHAPTER 4

60   *Evolutionary convergence:* A comprehensive and excellent account of
      convergence is found in Conway-Morris S. (2004) *Life's Solution: In-*
      evitable Humans in a Lonely Universe. Cambridge University Press,
      Cambridge. An erudite review can also be found in McGhee G. (2011)
      Convergent Evolution: Limited Forms Most Beautiful. Massachusetts
      Institute of Technology, Cambridge, MA. For a discussion of conver-
      gence, its potential confusion with other mechanisms of similarity,
      and the question of whether some instances of convergence are not
      truly independent but might emerge from the same developmental
      possibilities and phylogenetic constraints, particularly in closely re-
      lated organisms, see Wray GA. (2002) Do convergent developmental
      mechanisms underlie convergent phenotypes? *Brain, Behavior and*
      Evolution **59**, 327–336.

62   *This is the same with convergent evolution:* Mayr E. (2004) *What Makes*
      Biology Unique? Cambridge University Press, Cambridge, 71, states
      that "the physicochemical approach is totally sterile in evolutionary bi-
      ology. The historical aspects of biological organization are entirely out
      of reach of physicochemical reductionism." I agree with Mayr that con-
      tingency is not easy to reduce to equations, but I disagree that evolution
      is only historical. Many examples of convergence are underpinned by
      physical constraints and therefore do allow us to bring physicochemi-
      cal explanations forcefully into the evolutionary synthesis.

64   *Yet the question they ask:* Another question is why large animals, such
      as mammals and reptiles, have four legs and insects and other arthro-
      pods have six (or many more). The canonical response is that here
      at last we have an example of contingency at work, the number of
      limbs reflecting past body designs from which evolution cannot es-
      cape. In the case of arthropods, the segmented body plan allows for
      many pairs of limbs to be added or removed, leading to a range of
      numbers of legs among arthropods, from six-legged insects to ani-
      mals with just over seven hundred legs (e.g., the millipede *Illacme
      plenipes*). In vertebrates, ancestors with two sets of pectoral and pelvic
      fins led to a four-legged design. This theory of contingency for legs
      may well be the case—we could imagine, for instance, that a mammal
      version of a praying mantis that could run on four legs and have two
      appendages left free for predation might be very successful. More bio-
      mechanical studies may help discover to what extent leg number af-
      fects locomotion and whether, aside from contingency, there are any
      physical reasons for limb-number choices. See, however, Full RJ, Tu
      MS. (1990) Mechanics of six-legged runners. *Journal of Experimental
      Biology* **148**, 129–146, which suggests that the energy needed to move
      a center of mass a given distance is unchanged across many animals
      with different leg numbers.

64   *So why has nature:* LaBarbera M. (1983) Why the wheels won't go.
      American Naturalist **121**, 395–408.

66   *Richard Dawkins:* Dawkins R. (1996) Why don't animals have wheels?
      (London) Sunday Times, November 24.

66   *There is no selection pressure:* I note anecdotally tales of animals such
      as wolves using roads to get across forests more easily. Once these
      structures are built by an intelligent life form, other life may find
      roads useful for locomotion.

67   *Compare this:* I simplify somewhat. The menu of possibilities that fish
      and other aquatic animals use for locomotion is varied and impressive.
      Many fish, such as tuna, rely more on tail movements (thunniform
      swimming) than on body waves, and in this sense, their propulsion is
      closer to a propeller's.

67   *Dangling from the sides:* Berg HC, Anderson RA (1973) Bacteria swim
      by rotating their flagellar filaments. *Nature* **245**, 380–382 and Berg
      HC. (1974) Dynamic properties of bacterial flagellar motors. *Nature*
      249, 77–79.

68   *Despite the superficial similarities:* A rather famous paper that de-
      scribes these two very different worlds of low and high Reynolds

numbers is Purcell EM. (1977) Life at low Reynolds Number. *American Journal of Physics* **45**, 3–11.

69 *However, we find:* We could imagine wheels being useful on a planet dominated by smooth surfaces. However, even a planet without plate tectonics to create mountain ranges, such as Mars, has an irregular surface that would suit legged animals better than wheeled ones. A flat surface at the macroscopic scale does not change the irregular structure of a planet at the millimeter scale. As for propellers, there are other physical reasons for doubting that bacterial-like flagella motors can simply be scaled up. In a delightful and elegant essay on the topic of wheels, Gould S. (1983) Kingdoms without wheels. In *Hen's Teeth and Horse's Toes*, 158–165, suggests that flagella evolved and larger rotating structures did not because the bacterial apparatus depends on diffusion, which is too slow at large scales to enable the evolution of an analogous, but larger structure. Gould's argument suggests a physical barrier to explain something evolutionary, which in itself is interesting since he was fond of relegating physical processes to being an irrelevance against contingency.

69 *In 1917, a brilliant:* Thompson DW. (1992) *On Growth and Form.* Cambridge University Press, Cambridge.

69 *He explores the shapes:* The mathematical relationships in the growth of plants and their physical and biological bases have been a subject of fascination for centuries. Many plants exhibit growth that has a spiral pattern in leaves. You can even see it in pine cones. The arrangement of leaves in a plant is known as *phyllotaxis*. Often, two spirals—one clockwise and one anticlockwise—can be discerned in a plant leaf pattern if you look down toward its apex. Extraordinarily, the number of spirals running clockwise and anticlockwise is found to be two adjacent terms in the Fibonacci series. Each number in this special sequence is the sum of the preceding two numbers, so, for example, 1, 1, 2, 3, 5, 8, 13, 21, and so on, is a Fibonacci series. A plant might have 8 and 13 spirals. This arrangement is (8, 13) phyllotaxis. Why this should be the case is thought to be related to the packing of leaves to acquire maximum sunlight, to maximize efficiency of stacking, or both. This is reflected in the angle through which each leaf is rotated relative to the previous leaf. It is determined by physical principles, not a contingent product of natural selection. This intriguing mathematical relationship has long attracted the attention of physicists and mathematicians. Two examples to whet the appetite further: Newell AC, Shipman PD. (2005)

Plants and Fibonacci. *Journal of Statistical Physics* **121**, 927–968; and Mitchison GJ. (1977) Phyllotaxis and the Fibonacci series. *Science* **196**, 270–275. A classic paper offering a physical model is Douady S, Couder Y. (1991) Phyllotaxis as a physical self-organised growth response. *Physical Review Letters* **68**, 2098–2101. The link between Fibonacci series and biology is another beautiful example of the connection between biological form and predictable patterns in which the action of genes is to impose shape, structure, color, and other characteristics on an otherwise physically determined process.

72   *In his wonderful book:* Carroll SB. (2005) *Endless Forms Most Beautiful.* Quercus, London. He also explores the fascinating way in which wings in vertebrates developed in different ways from their limbs and digits. The transition between gills and wings in insects also shows the quite astonishing adaptability of the modules of animals to shape-shift into entirely new structures (Averof M, Cohen SM. [1997] Evolutionary origin of insect wings from ancestral gills. *Nature* **385**, 627–630).

73   *Some of this variation:* I do not discuss it here, but recent research on the development of flight in birds shows how specific gene regulatory elements are associated with the development of wings and feathers, another extraordinary example of how modest changes in genetic regulation and units can drive the acquisition of characteristics that allow organisms to tap into physical laws, in this case the rules of aerodynamics. For example, see Seki R et al. (2016) Functional roles of Aves class-specific cis-regulatory elements on macroevolution of bird-specific features. *Nature Communications* **8**, 14229.

74   *One transition is impressive:* Denny MW. (1993) *Air and Water: The Biology and Physics of Life's Media.* Princeton University Press, Princeton, NJ, doesn't explicitly address the move from water to land, but his treatise examining the physics of biology in water and in air, and often comparing both media and the implications for the structure of biological systems, is an impressive work. In many ways, it is one of the most detailed and ambitious paeans to the link between biology and physics to be written. Denny also explored aspects of life at the air-water interface. See, for example, Denny MW. (1999) Are there mechanical limits to size in wave-swept organisms? *Journal of Experimental Biology* **202**, 3463–3467.

74   *But the complete transition:* A very cogent account of this transition in the context of locomotion is given in Wilkinson M. (2016) *Restless Creatures: The Story of Life in Ten Movements.* Icon Books, London.

75   *The scientists found:* A series of papers document these insights. Just some worth reading are Freitas R, Zhang G, Cohn MJ. (2007) Biphasic *Hoxd* gene expression in shark paired fins reveals an ancient origin of the distal limb domain. *PLoS One* **8**, e754; Davis MC, Dahn RD, Shubin NH. (2007) An autopodial-like pattern of Hox expression in the fins of a basal actinopterygian fish. *Nature* **447**, 473–477; Schneider I et al. (2011) Appendage expression driven by the *Hoxd* Global Control Region is an ancient gnathosome feature. *Proceedings of the National Academy of Sciences* **108**, 12782–12786; Freitas R et al. (2012) Hoxd13 contribution to the evolution of vertebrate appendages. *Developmental Cell* **23**, 1219–1229; Davis MC. (2013) The deep homology of the autopod: Insights from Hox gene regulation. *Integrative and Comparative Biology* **53**, 224–232.

77   *The slightly more graceful:* A review of these different forms of movement is found in Gibb AC, Ashley-Ross MA, Hsieh ST. (2013) Thrash, flip, or jump: The behavioural and functional continuum of terrestrial locomotion in teleost fishes. *Integrative and Comparative Biology* **53**, 295–306. This paper was part of a wider symposium, and the review paper on the whole meeting is worth reading for an insight into the general problem that faced life in the transition from water to land: Ashley-Ross MA, Hsieh ST, Gibb AC, Blob RW. (2013) Vertebrate land invasions—past, present, and future: An introduction to the symposium. *Integrative and Comparative Biology* **53**, 1–5.

77   *Nonetheless, in the history of animal life:* Of course, arthropods also made this transition, giving rise to insects.

79   *Not surprisingly:* Cockell CS, Knowland J. (1999) Ultraviolet radiation screening compounds. *Biological Reviews* **74**, 311–345.

80   *In experiments that tracked:* A fascinating paper that describes these findings is Leal F, Cohn MJ. (2016) Loss and re-emergence of legs in snakes by modular evolution of *Sonic hedgehog* and HOXD enhancers. *Current Biology* **26**, 1–8.

80   *Almost certainly, the secrets:* A very readable book describing the history of research investigating the invasion of land and the return to water is Zimmer C. (1998) *At the Water's Edge.* Touchstone, New York.

81   *Charles Darwin concluded:* Darwin C. (1859) *On the Origin of Species by Means of Natural Selection, or the Preservation of Favoured Races in the Struggle for Life.* John Murray, London.

CHAPTER 5

83   *However, if I tell you:* Bianconi E et al. (2013) An estimation of the number of cells in the human body. *Annals of Human Biology* **40**, 463–471.

84   *Hooke had no inkling:* Hooke R. (1665) *Micrographia.* J Martyn and J Allestry, printers to the Royal Society, London.

84   *What he found:* Van Leeuwenhoek published a prodigious number of letters on many things he observed under his microscopes, not merely his animalcules. One seminal paper on observations about microbes is Leeuwenhoek A. (1677) Observation, communicated to the publisher by Mr. Antony van Leuwenhoek, in a Dutch letter of the 9 Octob. 1676 here English'd: concerning little animals by him observed in rain-well-sea and snow water; as also in water wherein pepper had lain infused. *Philosophical Transactions* **12**, 821–831.

86   *Viruses, small pieces:* Viruses contain either DNA or RNA, and these molecules can be in either single- or double-stranded form. For protein coats, note that the assembly of these relatively simple entities can be understood in physical terms. A classic paper describing the geometry of viruses and the mathematical and physical principles that shape their protein coats is Caspar DLD, Klug A. (1962) Physical principles in the construction of regular viruses. *Cold Spring Harbor Symposia on Quantitative Biology* **27**, 1–24.

88   *The first molecules:* In this book, I do not address whether the emergence of life is inevitable. This omission is not a capitulation. I regard the issue as a different question. We do not know whether life is inevitable on a planet with water and clement conditions. It is a profound question at the interface between biology and physics to ask whether suitable physical conditions on a rocky planet will inevitably lead to life. I am interested here in the restrictions on life once it does emerge. However, an intriguing take on the physical and chemical basis of the origin of life is pursued by Pross A. (2012) *What Is Life?* Oxford University Press, Oxford.

88   *the 1980s, David Deamer:* A detailed account of this work can be found in Deamer D. (2011) *First Life: Discovering the Connections Between Stars, Cells, and How Life Began.* University of California Press, Berkeley. In this book, Deamer also explores many other conundrums about the origin of life as well as how cellularity allowed

for complexity of metabolic processes. A paper summarizing results with self-assembling membranes is Deamer D et al. (2002) The first cell membranes. *Astrobiology* **2**, 371–381.

89   *Deamer had shown:* A paper that takes a theoretical approach to understanding the physics of self-assembling vesicles and even their reproduction is Svetina S. (2009) Vesicle budding and the origin of cellular life. *ChemPhysChem* **10**, 2769–2776.

89   *Provided that the gases:* One of the strangest links between membranes and astronomy is the observation that in the endoplasmic reticulum (the organelle responsible for protein synthesis in eukaryotic cells), layers of membranes are attached to one another in stacks linked with helicoid ramps in a shape resembling a parking garage. Similar structures are thought to exist in the extreme conditions of neutron stars. Whether this similarity of shape is coincidence or reflects some underlying physical principle to do with energy minimization is not known, but the bizarre observation might reflect the common physics underlying patterns in nature: Berry DK et al. (2016) "Parking-garage" structures in nuclear astrophysics and cellular biophysics. *Physical Review* C **94**, 055801.

89   *Pyruvic acid:* Described in Deamer (2011) *First Life*, above.

89   *Although these ingenious:* Some ideas on the environments and processes leading to the first protocells are nicely described in Black RA, Blosser MC. (2016) A self-assembled aggregate composed of a fatty acid membrane and the building blocks of biological polymers provides a first step in the emergence of protocells. *Life* **6**, 33.

90   *Some scientists think:* Martin W, Russell MJ. (2007) On the origin of biochemistry at an alkaline hydrothermal vent. *Philosophical Transactions of the Royal Society* **362**, 1887–1926.

90   *Others think:* Cockell CS. (2006) The origin and emergence of life under impact bombardment. *Philosophical Transactions of the Royal Society* **1474**, 1845–1855.

90   *Perhaps Darwin's "warm little pond":* Darwin described the origin of life in a letter to his friend Joseph Hooker (February 1, 1871): "But if (and oh what a big if) we could conceive in some warm little pond with all sorts of ammonia & phosphoric salts,—light, heat, electricity etc present, that a protein compound was chemically formed, ready to undergo still more complex changes, at the present day such matter would be instantly devoured, or absorbed, which would not have been the case before living creatures were formed."

90   *Deamer's experiments . . . schizophrenic molecules:* More technically termed *amphiphilic.*

91   *The simplicity of these pathways:* Smith E, Morowitz HJ. (2004) Universality in intermediary metabolism. *Proceedings of the National Academy of Sciences* **101**, 13,168–13,173.

91   *By testing thousands:* Court SJ, Waclaw B, Allen RJ. (2015) Lower glycolysis carries a higher flux than any biochemically possible alternative. *Nature Communications* **6**, 8427. However, the authors did also show that the route nature uses is not the only possibility. Under different environmental conditions in the cell, other pathways could be used.

91   *These independent investigations:* Similar findings of strong selection based on physical considerations have been reported in studies of the regulatory networks involved in the cell cycle. A high degree of robustness to perturbation was found. See, for example, Li F, Long T, Lu Y, Ouyang Q, Tang C. (2004). The yeast cell-cycle network is robustly designed. *Proceedings of the National Academy of Sciences* **101**, 4781–4786. The information within biological networks may be different from purely random networks and may provide ways of understanding which physical principles are instantiated into biological systems as they emerge. See Walker SI, Kim H, Davies PCW. (2016) The informational architecture of the cell. *Philosophical Transactions of the Royal Society A: Mathematical, Physical and Engineering Sciences* **374**, article 0057.

92   *Nor do these sorts of conclusions:* Computational modeling has been used to study how easily one metabolic pathway can change into another and thus whether existing pathways are very much a product of historical quirks. Barve and colleagues conclude their paper with this comment about flexibility of pathways: "Metabolism is thus highly evolvable. . . . Historical contingency does not strongly restrict the origin of novel metabolic phenotypes." Barve A, Hosseini S-R, Martin OC, Wagner A. (2014) Historical contingency and the gradual evolution of metabolic properties in central carbon and genome-scale metabolism. *BMC Systems Biology* **8**, 48.

92   *Instead, many metabolic process:* The pervasive question of predictability in evolutionary possibilities has received some attention at the scale of biochemical pathways. At the metabolic level, with knowledge of an organism's environment and lifestyle, we can apparently predict with quite surprising accuracy where and how a pathway will develop (Pál C et al. [2006] Chance and necessity in the evolution of

minimal metabolic networks. *Nature* **440**, 667–670). An important paper suggests that only a few designs, or *topologies*, of biochemical pathways are possible. This work suggests that at the very least, the structure of biochemical networks may be quite predictable (Ma W et al. [2009] Defining network topologies that can achieve biochemical adaptation. *Cell* **138**, 760–773).

92   *When the first cells moved:* How the environment might fashion the shape of microbes is beautifully described in Young KD. (2006) The selective value of bacterial shape. *Microbiology and Molecular Biology Reviews* **70**, 660–703.

92   *One first indefatigable effect:* For a discussion on the ethical implications of a hypothetical world in which bacteria are the size of dogs, see Cockell CS. (2008) Environmental ethics and size. *Ethics and the Environment* **13**, 23–39.

92   *There are many causes:* Several essays exploring a range of factors influential in determining cell size are found in: Marshall WF et al. (2012) What determines cell size? *BMC Biology* 10.101. An excellent and simple discussion of the role of diffusion is Vogel S. (1988) *Life's Devices: The Physical World of Animals and Plants.* Princeton University Press, Princeton, NJ.

92   *A large bag:* A fascinating study has suggested that if cells grow larger than about ten micrometers, gravity begins to become significant and may explain why large frog ovary cells, which can be greater than one millimeter in diameter, have a molecular (actin) scaffold around their nucleus to stabilize them against the effects of gravity (Feric M, Brangwynne CP. [2013] A nuclear F-actin scaffold stabilizes ribonucleoprotein droplets against gravity in large cells. *Nature Cell Biology* **15**, 1253–1259). An interesting implication is the speculation that on planets with lower gravity, larger cells may be possible, all other things being equal.

93   *The larger you are:* Beveridge TJ. (1988) The bacterial surface: General considerations towards design and function. *Canadian Journal of Microbiology* **34**, 363–372. Diffusion may be less important a factor than was once supposed. The cell interior turns out to be remarkably crowded, and the model of a molecule diffusing passively through a fluid is too simple.

93   *That smallest theoretical size:* This size limit was arrived at by a group of people attempting to define the smallest expected microbe. They were partly motivated to ascertain what the smallest possible biological

signature of a cell might be on another planet, for example on Mars. The study is intriguing, as it was provoked by researchers' desire to set a universal boundary of cell size (National Research Council Space Studies Board [1999] *Size Limits of Very Small Microorganisms*. National Academies Press, Washington, DC. However, note that a size range of one hundred to three hundred nanometers for the minimum cell was arrived at by Alexander RM. (1985) The ideal and the feasible: Physical constraints on evolution. *Biological Journal of the Linnean Society* **26**, 345–358.

93   *The estimate actually fits:* Do not be fooled. This diminutive creature can reach up to 50 percent of all the biomass in surface water in the oceans. It is immensely important in cycling carbon in the Earth's oceans.

94   *When we talk of smallness:* Descriptions of these microbes and some discussion of the physics behind their lifestyles are to be found in the very clearly titled paper Schulz HN, Jørgensen BB. (2001) Big bacteria. *Annual Reviews of Microbiology* **55**, 105–137.

95   *Here, instead of nutrient needs:* Described in Persat A, Stone HA, Gitai Z. (2014) The curved shape of *Caulobacter crescentus* enhances surface colonization in flow. *Nature Communications* **5**, 3824.

95   *In these places:* Kaiser GE, Doetsch RN. (1975) Enhanced translational motion of *Leptospira* in viscous environments. *Nature* **255**, 656–657.

96   *These laws:* In this context, I refer the reader to fascinating work by Jeremy England and his group, who suggest that adaptation can be realized without selection. Chemical systems can fine-tune their processes in response to their environments as the system establishes resonances with the very environmental factors acting on it. This observation should not be taken as yet another hackneyed attempt to prove that Darwin was wrong. Instead, it shows that organic matter's ability to evolve may well be aided by its natural tendency to take on forms that reflect its environment, even before the environment has acted to select the forms of that matter that successfully reproduce. England and his colleagues' work show that biological evolution does not work unexpectedly against disorder, but that emergent complexity in physical systems, including life, favors this process. See, for example, Horowitz JM, England JL. (2017) Spontaneous fine-tuning to environment in many-species chemical reaction networks. *Proceedings of the National Academy of Sciences* **114**, 7565–7570; and Kachman T, Owen JA, England JL. (2017) Self-organized resonance during search of a diverse chemical space. *Physical Review Letters* **119**, 038001.

96   *In this medley:* In some outstanding experiments, Richard Lenski's
      group studied populations of the bacterium *Escherichia coli* to see
      if they could separate the effects of adaptation, chance, and histori-
      cal influence in microbial evolution. They found that adaptation was
      extraordinarily versatile, allowing organisms to mutate to achieve a
      similar fitness with little effect of contingency or history. However, in
      traits that were not so important for fitness in these particular exper-
      iments, such as cell size (which may, however, be important in more
      natural environments), contingency could throw up variants presum-
      ably because the effects of these mutants were neutral. History could
      also affect subsequent cell size. Their observations are probably quite
      generalizable; if a trait has little direct impact on survival to reproduc-
      tive age in any organisms, then the trait may be more susceptible to
      chance alterations or it may reflect the idiosyncrasies of past historical
      attributes. Travisano M, Mongold JA, Bennett AF, Lenski RE. (1995)
      Experimental tests of the roles of adaptation, chance, and history in
      evolution. *Science* **267**, 87–89.

97   *Once the prey:* Proposed by Lake JA. (2009) Evidence for an early pro-
      karyotic endosymbiosis. *Nature* **460**, 967–971.

98   *These chemical products:* This alternative idea was suggested by Gupta
      RS. (2011) Origin of the diderm (Gram-negative) bacteria: Antibiotic
      selection pressure rather than endosymbiosis likely led to the evolu-
      tion of bacterial cells with two membranes. *Antonie van Leeuwenhoek*
      100, 171–182.

98   *In the archaea:* The charged heads of the lipids that stick into the wa-
      ter are linked to their long chains by ether linkages in the archaea,
      rather than the more familiar ester linkages in bacteria. For an in-
      depth discussion of their chemical differences, see Albers S-V, Meyer
      BH. (2011) The archaeal cell envelope. *Nature Reviews Microbiology*
      9, 414–426.

100   *Cooperation, forced:* A classic paper presenting a view of the multi-
      cellular capacities of bacteria is Shapiro JA. (1998) Thinking about
      bacterial populations as multicellular organisms. *Annual Reviews of
      Microbiology* **52**, 81–104. Another view is Aguilar C, Vlamakis H,
      Losick R, Kolter R. (2007) Thinking about *Bacillus subtilis* as a multi-
      cellular organism. *Current Opinion in Microbiology* **10**, 638–643.

100   *These patterns and order arise:* Like ants, birds, and schooling fish
      (Chapter 2), bacteria are the focus of physicists studying active mat-
      ter. Their collective behavior lends itself to modeling and simulation.

See Copeland MF, Weibel DB. (2009) Bacterial swarming: A model system for studying dynamic self-assembly. *Soft Matter* **5**, 1174–1187; and Wilking JN et al. (2011) Biofilms as complex fluids. *Materials Research Society (MRS) Bulletin* **36**, 385–391.

100 *Equations can be used:* The responses of large numbers of cells to chemical cues can be modeled. See, for example, Camley BA, Zimmermann J, Levine H, Rappel W-J. (2016) Emergent collective chemotaxis without single-cell gradient sensing. *Physical Review Letters* **116**, 098101.

101 *domain of the eukaryotes: Prokaryote*, literally translated as "before the kernel," encompasses microbes without a cell nucleus (i.e., most microbes on Earth), in contrast to the eukaryotes ("true nucleus"), organisms whose cells generally contain a nucleus. Eukaryotes do include some single-celled microbes such as algae and some fungi, including yeasts, but these single-celled organisms have a nucleus and other organelles.

101 *The eukaryotic cell is:* It is established that endosymbiosis led to the chloroplast, the photosynthetic apparatus in algae and plants. It began its life as an engulfed cyanobacterium.

101 *Many hundreds:* Lane N, Martin W. (2010) The energetics of genome complexity. *Nature* **467**, 929–934.

101 *Coupled with this:* The potential role of genome complexity as another major difference between prokaryotes and eukaryotes as a critical pathway to animal life is elaborated on in Lynch M, Conery JS. (2003) The origins of genome complexity. *Science* **302**, 1401–1404.

102 *Microbes that could grab:* Photosynthesis using oxygen evolved only once: the early cyanobacteria that mastered this trick eventually became engulfed to make algae and plants. Although photosynthesis has appeared once, it is not necessarily an unlikely evolutionary development, a contingent fluke. Instead, once that feat had been achieved, habitats became filled with photosynthesizers and there may have been few niches left for a second evolution of this pathway to move into.

103 *Endosymbiosis has happened:* Just one such example is Marin BM, Nowack EC, Melkonian M. (2005) A plastid in the making: Evidence for a second primary endosymbiosis. *Protist* **156**, 425–432.

103 *Then the cells gather:* Slime molds can even be used to re-create the most efficient connections between two points. By placing them on maps in the laboratory (where cities and towns are globs of food),

they can even be used to predict the best road and rail networks across landscapes such as the Tokyo rail system (Tero A et al. [2010] Rules for biologically inspired adaptive network design. *Science* **327**, 439–442) or Brazilian highways (Adamatsky A, de Oliveira PPB. [2011] Brazilian highways from slime mold's point of view. *Kybernetes* **40**, 1373–1394). Many other countries' transport networks have been scrutinized using *Physarum plasmodium* and *P. polycephalum*.

103   *We do not know the exact events:* There are many theories for how this might have happened. Insights into cell communication and the genetics of how cells attach and signal are likely to reveal many steps that led from unicellular organisms to true multicellular (differentiated) organisms. See, for example, King N. (2004) The unicellular ancestry of animal development. *Developmental Cell* **7**, 313–325; and Richter DJ, King N. (2013) The genomic and cellular foundations of animal origins. *Annual Reviews of Genetics* **47**, 509–537.

103   *Put simply, the rise of multicellularity:* Multicellular structures may emerge from the interaction of physical principles. See Newman SA, Forgacs G, Müller GB. (2006) Before programs: The physical origination of multicellular forms. *International Journal of Developmental Biology* **50**, 289–299.

103   *A biological arms race:* Dawkins R, Krebs JR. (1979) Arms races between and within species. *Proceedings of the Royal Society* **205**, 489–511. Regarding sometimes larger machines, a tendency to become larger may also occur simply because organisms are constrained by the minimum size of cells in becoming smaller so they inevitably move into the larger morphospace of possible forms. See Gould SJ. (1988) Trends as changes in variance: A new slant on progress and directionality in evolution. *Journal of Paleontology* **62**, 319–329. Even this process, though, results from a simple physical principle: organisms become larger to fill available niches that will allow for larger organisms (e.g., on account of energy available for such forms).

104   *The second claim:* We must also be mindful that a planet has a finite lifetime. If the stages along the way between a microbe and a mammoth exceed the time that a planet hosts habitable conditions, then the experiment in evolution will be cut short. This sad end is rooted unambiguously in physics too, the evolution of a star unkindly intercepting the trajectory of life.

104   *Considering the inevitability:* It has been proposed that very few, if any, of the innovations between the origin of life and the numerous

key adaptations in multicellular organisms may be unique, that is, singularities in the evolutionary process. See, for example, Vermeij GJ. (2006) Historical contingency and the purported uniqueness of evolutionary innovations. *Proceedings of the National Academy of Sciences* **103**, 1804–1809.

CHAPTER 6

105   *Jump in a car:* Woods PJE. (1979) The geology of Boulby mine. *Economic Geology* **74**, 409–418.

107   *From the science-fiction cleanliness:* The laboratory is run by Sean Paling and his team. Many people need to be thanked, including Emma Meehan, Lou Yeoman, Christopher Toth, Barbara Suckling, Tom Edwards, Jac Genis, David McLuckie, David Pybus, and others who have made our work at Boulby possible.

108   *Here, animal life:* Two nice general books on the extremophiles are Gross M. (2001) *Life on the Edge: Amazing Creatures Thriving in Extreme Environments.* Basic Books, New York; and Postgate JR. (1995) *The Outer Reaches of Life.* Cambridge University Press, Cambridge.

109   *Life deep down:* An excellent book that provides an insight into the history and science of deep subsurface life is Onstott TC. (2017) *Deep Life: The Hunt for the Hidden Biology of Earth, Mars, and Beyond.* Princeton University Press, Princeton, NJ.

110   *Within the unremarkable sludge:* Brock TD, Hudson F. (1969) *Thermus aquaticus* gen. n. and sp. n., a nonsporulating extreme thermophile. *Journal of Bacteriology* **98**, 289–297.

110   *Among their ranks:* Takai K et al. (2008) Cell proliferation at 122 °C and isotopically heavy CH_4 production by a hyperthermophilic methanogen under high-pressure cultivation. *Proceedings of the National Academy of Sciences USA.* **105**, 10949–10954.

110   *Proteins can be made:* For a significant paper that shows how the adaptations of proteins to high temperatures can be explained in terms of physical principles, see Berezovsky IN, Shakhnovich EI. (2005) Physics and evolution of thermophilic adaptation. *Proceedings of the National Academy of Sciences* **102**, 12,742–12,747.

111   *One group of researchers:* Cowan DA. (2004) The upper temperature for life—where do we draw the line? *Trends in Microbiology* **12**, 58–60.

111   *At temperatures of around 450°C:* For the upper temperature limits of
      life as set by molecular stability, see Daniel RM, Cowan DA. (2000)
      Biomolecular stability and life at high temperatures. *Cellular and
      Molecular Life Sciences* **57**, 250–264.

112   *Life on Earth should:* Cockell CS. (2011) Life in the lithosphere, kinet-
      ics and the prospects for life elsewhere. *Philosophical Transactions of
      the Royal Society* **369**, 516–537.

113   *So far, there is no good:* A paper quantifying lower temperature limits
      for life is Price PB, Sowers T. (2004) Temperature dependence of met-
      abolic rates for microbial growth, maintenance, and survival. *Proceed-
      ings of the National Academy of Sciences* **101**, 4631–4636. At very low
      temperatures, cells reach a point when the rate of energy expenditure
      by microbes is only just able to keep up with the rate of damage. This
      trade-off will ultimately determine the lower temperature limit for
      any given life form to remain intact over long periods.

113   *The challenge that low-temperature life:* Another paper that examines
      this problem considers the challenges caused by liquids that "vitrify,"
      essentially turn into a glasslike state in cells at low temperatures. Vit-
      rification is likely to seriously limit the movement of gases and nutri-
      ents and may set a lower limit for life in many organisms. See Clarke
      A et al. (2013) A low temperature limit for life on Earth. *PLoS One* **8**,
      e66207.

113   *This* background radiation: The radiation may not be entirely
      detrimental. The radiolysis of water, or its breaking up by radi-
      ation, could release hydrogen, which microbes can use as an en-
      ergy source. See, for example, Lin L-H et al. (2005) Radiolytic H_2
      in continental crust: Nuclear power for deep subsurface microbial
      communities. *Geochemistry, Geophysics and Geosystems* **6**, doi:
      10.1029/2004GC000907.

114   *Added to these problems:* An example is depurination in the DNA,
      in which the β-N-glycosidic bond is hydrolytically cleaved, releas-
      ing a nucleic base, adenine or guanine, from the DNA structure. See
      Lindahl T. (1993) Instability and decay of the primary structure of
      DNA. *Nature* **362**, 709–715; Lindahl T, Nyberg B. (1972) Rate of depu-
      rination of native deoxyribonucleic acid. *Biochemistry* **11**, 3610–3618.

114   *After eating your breakfast:* Lipids that make up membranes contain
      fatty acids, long chains of carbon atoms. When we talk about fats in
      butter, we are talking about the same material—fatty acids.

115   *When exposed to subfreezing:* A review that summarizes the variety of challenges and solutions for life at low temperatures is D'Amico S et al. (2006) Psychrophilic microorganisms: Challenges for life. *EMBO Reports* 7, 385–389.

115   *In this zone, reactions:* Including the process of racemization, the tendency for the chirality (L- or D- forms of molecules) to be lost. Remember that amino acids in life are primarily in the L-form. Racemization will tend to produce an equal amount of L- and D-forms. It can happen inexorably over time by thermal effects on molecules. The racemization of amino acids and low-temperature environments is discussed in Brinton KLF, Tsapin AI, Gilichinsky D, McDonald GD. (2002) Aspartic acid racemization and age-depth relationships for organic carbon in Siberian permafrost. *Astrobiology* 2, 77–82.

116   *No one should be surprised:* Grant S et al. (1999) Novel archaeal phylotypes from an East African alkaline saltern. *Extremophiles* 3, 139–145.

117   *Faced with the trauma:* The problems of high salt are described in Oren A. (2008) Microbial life at high salt concentrations: Phylogenetic and metabolic diversity. *Saline Systems* 4, doi: 10.1186/1746-1448-4-2. For the thermodynamic limitations imposed by salt, see Oren A. (2011) Thermodynamic limits to microbial life at high salt concentrations. *Environmental Microbiology* 13, 1908–1923.

118   *Like scientists enamored:* Stevenson A et al. (2015) Is there a common water-activity limit for the three domains of life? *ISME J* 9, 1333–1351.

118   *By a water activity:* Stevenson A et al. (2017) *Aspergillus penicillioides* differentiation and cell division at 0.585 water activity. *Environmental Microbiology* 19, 687–697.

119   *These solutions can also cause disorder:* Hallsworth JE et al. (2007) Limits of life in $MgCl_2$-containing environments: Chaotropicity defines the window. *Environmental Microbiology* 9, 801–813.

119   *When investigated by microbiologists:* Yakimov MM et al. (2015) Microbial community of the deep-sea brine Lake Kryos seawater–brine interface is active below the chaotropicity limit of life as revealed by recovery of mRNA. *Environmental Microbiology* 17, 364–382.

119   *Indeed, microbiologists have had mixed results:* Siegel BZ. (1979) Life in the calcium chloride environment of Don Juan Pond, Antarctica. *Nature* 280, 828–829.

120   *Yet we find life thriving:* Amaral Zettler LA et al. (2002) Microbiology: Eukaryotic diversity in Spain's River of Fire. *Nature* **417**, 137.

121   *The acid-loving microbes:* The adaptations to low pH are summarized well in Baker-Austin C, Dopson M. (2007) Life in acid: pH homeostasis in acidophiles. *Trends in Microbiology* **15**, 165–171. For insights into adaptations from the genome, see Ciaramella M, Napoli A, Rossi M. (2005) Another extreme genome: How to live at pH 0. *Trends in Microbiology* **13**, 49–51.

121   *A trip to Mono Lake:* Humayoun SB, Bano N, Hollibaugh JT. (2003) Depth distribution of microbial diversity in Mono Lake, a meromictic soda lake in California. *Applied and Environmental Microbiology* **69**, 1030–1042.

122   *In most of Earth's environments:* Some nice work was done by Jesse Harrison, a postdoctoral scientist in my laboratory, to attempt to map the limits of life using growth ranges of known strains of bacteria in the laboratory. You end up with intriguing three-dimensional plots of the boundary space of life: Harrison JP, Gheeraert N, Tsigelnitskiy D, Cockell CS. (2013) The limits for life under multiple extremes. *Trends in Microbiology* **21**, 204–212. This work used only laboratory strains, but natural environments outside the extremes in this paper are known to contain microbes, so there is still much to do to define the physical and chemical boundary space of life.

122   *Microbes have been found:* Mesbah NM, Wiegel J. (2008) Life at extreme limits: The anaerobic halophilic alkalithermophiles. *Annals of the New York Academy Sciences* **1125**, 44–57.

122   *Other extremes too:* Oger PM, Jebbar M. (2010) The many ways of coping with pressure. *Research in Microbiology* **161**, 799–809.

122   *Pores and transporters:* Bartlett DH. (2002) Pressure effects on in vivo microbial processes. *Biochimica et Biophysica Acta* **1595**, 367–381.

123   *The humble* Chroococcidiopsis: Billi D et al. (2000) Ionizing-radiation resistance in the desiccation-tolerant cyanobacterium *Chroococcidiopsis. Applied and Environmental Microbiology* **66**, 1489–1492.

123   *microbe joins:* Perhaps the most famous radiation-resistant microbe is *Deinococcus radiodurans* (a Greek and Latin portmanteau literally meaning "radiation-surviving terrible berry"). See Cox MM, Battista JR. (2005) *Deinococcus radiodurans*—the consummate survivor. *Nature Reviews Microbiology* **3**, 882–892. However, its capacities

are not unique. Other bacteria (including *Chroococcidiopsis* and *Rubrobacter*) also have high radiation tolerance.

124 *This zoo:* This observation doesn't contradict the fact that within the confines of the zoo, life is remarkably tenacious and can occupy a startling range of physical and chemical conditions. For a jaunt through these capacities and life's ability to ride out catastrophes that occur during its tenure on Earth, see Cockell CS. (2003) *Impossible Extinction: Natural Catastrophes and the Supremacy of the Microbial World.* Cambridge University Press, Cambridge.

CHAPTER 7

126 *In an early paper:* Crick FHC. (1965) The origin of the genetic code. *Journal of Molecular Biology* **38**, 367–379.

127 *This apparently odd property:* Watson JD, Crick FHC. (1953) A structure for deoxyribose nucleic acid. *Nature* **171**, 737–738.

127 *Surely it is just chance:* The number of "letters" in the genetic code is reviewed by Szathmáry E. (2003) Why are there four letters in the genetic code? *Nature Reviews in Genetics* **4**, 995–1001.

127 *In this "RNA world":* Higgs PG, Lehman N. (2015) The RNA World: Molecular cooperation at the origins of life. *Nature* **16**, 7–17

128 *We do not discover:* The reader might well retort that the argument is tautological: Of course the models give us results congruent with Earth's biology because the models we used are based on RNA, the very molecule that Earth life uses! I would reply with the very unscientific "maybe." However, as is apparent later in this chapter, we can explore many alternative base pairs and molecules, which suggest that the choice of chemicals in the genetic code is not chance. There are some genetic-code-like molecules that have similarities with the other classes of molecules that make life, for example PNA, peptide nucleic acids, which crudely have protein-like qualities with their peptide bonds. However, no one has yet shown that the other major monomers of life that are thought to have been present on the early Earth (e.g., amino acids, lipids, and sugars) can form a genetic code. Among the various organic molecules on offer for the first living things, the ones used in our genetic code seem likely. Nevertheless, we should be open-minded about possible alternative chemistries for genetic codes. I limit myself in this chapter to the observation

that once the nucleotides were evolutionarily selected as the basis of the genetic code, the rest of the architecture of the code and its molecular products is highly noncontingent and driven by physical considerations.

129 *Motivated by a desire:* Zhang Y et al. (2016) A semisynthetic organism engineered for the stable expansion of the genetic alphabet. *Proceedings of the National Academy of Sciences*, doi: 10.1073/pnas.1616443114

129 *The unwieldly named xanthosine:* Piccirilli JA et al. (1990) Enzymatic incorporation of a new base pair into DNA and RNA extends the genetic alphabet. *Nature* **343**, 33–37.

130 *Some isoguanine and isocytosine:* Malyshev DA et al. (2014) A semi-synthetic organism with an expanded genetic alphabet. *Nature* **509**, 385–388.

130 *Scientists at various institutions:* Reviewed in Eschenmoser A. (1999) Chemical etiology of nucleic acid structure. *Science* **284**, 2118–2124.

133 *Perhaps the organization of the amino acids:* Error minimization as a strong selection pressure for the code is described in a number of papers, for example Freeland SJ, Knight RD, Landweber LF, Hurst LD. (2000) Early fixation of an optimal genetic code. *Molecular Biology and Evolution* **17**, 511–518.

134 *Furthermore, amino acids:* Other factors have been proposed. For example, horizontal gene transfer (the movement of genes from one cell or organism to another) can increase the selection for optimal codes. See Sengupta S, Aggarwal N, Bandhu AV. (2014) Two perspectives on the origin of the standard genetic code. *Origins of Life and Evolution of Biospheres* **44**, 287–292.

134 *Of all the codes:* In an enticing Las Vegas–style titled paper, this analysis is described in Freeland SJ, Hurst LD. (1998) The genetic code is one in a million. *Journal of Molecular Evolution* **47**, 238–248.

134 *It is easy to get sucked into:* A critique of different pressures shaping the early code was made by Knight RD, Freeland SJ, Landweber LF. (1999) Selection, history and chemistry: The three faces of the genetic code. *Trends in Biochemical Sciences* **24**, 241–247, who suggest that different pressures may have dominated at different stages in the origin and early evolution of life. The pathways to the code and the role of coevolution in the process is also discussed by Wong, JT-F et al. (2016) Coevolution theory of the genetic code at age forty: Pathway

to translation and synthetic life. *Life* **6**, doi: 10.3390/life6010012. Another excellent review of the problems is Koonin EV, Novozhilov AS. (2009) Origin and evolution of the genetic code: The universal enigma. *Life* **61**, 99–111.

136 *Like much about biology:* Schrödinger had a pretty good crack at it. See Schrödinger E. (1944) *What Is Life?* Cambridge University Press, Cambridge.

137 *Churning out from the RNA:* Biological catalysts, or enzymes, perform a vast number of chemical reactions in cells and speed them up to much higher rates than would happen without them.

137 *Curious researchers have long wondered:* Amino acids, like many chemicals, come in two types: left-handed (L-amino acids) and right-handed (D-amino acids). In analogy to your two hands, these two forms are mirror images of each other. The left- and right-handed forms rotate polarized light either anticlockwise (to the left, or levorotation) or clockwise (to the right, or dextrorotation), respectively, hence their designation as L- or D-forms. Almost all amino acids in life (barring some in membranes, for instance) are L-amino acids. The preponderance of L-forms was thought to be a matter of chance in life, but some evidence suggests that amino acids in meteorites are partly enriched in L-forms (see, for example, Engel MH, Macko SA. [1997] Isotopic evidence for extraterrestrial non-racemic amino acids in the Murchison meteorite. *Nature* **389**, 265–268), suggesting an enrichment of the L-form of these amino acids in prebiotic molecules used by life. An alternative explanation is that polarized light in interstellar clouds preferentially destroyed one chiral form over the other, leading to initial enrichment of chiral molecules, later used in prebiotic chemistry (Bonner WA. [1995] Chirality and life. *Origins of Life and Evolution of Biospheres* **25**, 175–190). As life depends on molecular recognition and so is made simpler if all molecules are one form or the other, it is likely that the L-form may have been amplified until it became the predominant form. An interesting question is whether life elsewhere, if it exists, could be made of either L- or D-amino acids. The question strikes at the heart of the basic question of whether contingency or physics drove the early events of evolution. We might equally ask this question about the sugars. Our sugars are predominantly in the D-form.

137 *Initial attempts to discover:* An excellent study was Weber AL, Miller SL. (1981) Reasons for the occurrence of the twenty coded protein amino acids. *Journal of Molecular Evolution* **17**, 273–284.

137 *But then in 2011, Gayle Philip:* Philip GK, Freeland SJ. (2011) Did evolution select a nonrandom "alphabet" of amino acids? *Astrobiology* **11**, 235–240.

140 *In more recent years, synthetic biologists:* See, for example, Tiang Y, Tirrell DA. (2002) Attenuation of the editing activity of the *Escherichia coli* leucyl-tRNA synthetase allows incorporation of novel amino acids into proteins in vivo. *Biochemistry* **41**, 10,635–10,645.

140 *After all, if some of these new amino acids:* I refer here to natural selection. Humans are now making these changes artificially.

141 *The unusual amino acid:* Johansson L, Gafvelin G, Amér ESJ. (2005) Selenocysteine in proteins—properties and biotechnological use. *Biochimica et Biophysica Acta* **1726**, 1–13.

142 *Another strange cousin:* Srinivasan G, James CM, Krzycki JA. (2002). Pyrrolysine encoded by UAG in Archaea: Charging of a UAG-decoding specialized tRNA. *Science* **296**, 1459–1462.

142 *Yet as these molecules were uncoiled:* The implications of the limited protein folding possibilities for our understanding of evolution is nicely explored in Denton MJ, Marshall CJ, Legge M. (2002) The protein folds as platonic forms: New support for the pre-Darwinian conception of evolution by natural law. *Journal of Theoretical Biology* **219**, 325–342, which also discusses how this knowledge might imply the existence of laws of biology rooted in physical principles.

142 *Helices (termed α-helices):* A carboxyl group.

143 *As all these folds collapse:* Other factors, such as stability against mutations, may select for certain protein folds. Fascinating papers that explore the reasons for limited protein folds include Li H, Helling R, Tang C, Wingren N. (1996) Emergence of preferred structures in a simple model of protein folding. *Science* **273**, 666–669; and Li H, Tang C, Wingren N. (1998) Are protein folds atypical? *Proceedings of the National Academy of Sciences* **95**, 4987–4990. Weinreich and colleagues, in a study of the mutational trajectories of a bacterial protein, are quite explicit that "this implies that the protein tape of life may be largely reproducible and even predictable" (Weinreich DM, Delaney NF, DePristo MA, Hartl DL. [2006] Darwinian evolution can follow only very few mutational paths to fitter proteins. *Science* **312**, 111–113).

144 *Darwinian evolution:* Mutations and genetic shuffling and movement such as horizontal gene transfer generate an inexorable increase in diversity. This tendency has even been instantiated into a law (McShea DW, Brandon RN. [2010] *Biology's First Law: The Tendency for*

Diversity and Complexity to Increase in Evolutionary Systems. University of Chicago Press, Chicago). However, the degree to which such a tendency can really be a law or merely reflects the inexorable process of mutation that will occur in a code is a matter for debate. If any law drives this proposed biological phenomenon, it is probably the second law of thermodynamics.

CHAPTER 8

146 *The second law of thermodynamics:* Borgnakke C, Sonntag RE. (2009) *Fundamentals of Thermodynamics.* Wiley, Chichester.

147 *Sitting on cell membranes:* The mitochondria are the organelles that produce energy in most eukaryotic cells. The electron transfer chain that I describe occurs within the membranes of mitochondria. In prokaryotes, the transfer chain occurs in the cellular membrane, not in an organelle.

148 *Mitchell creatively figured:* Mitchell P. (1961) The chemiosmotic hypothesis. *Nature* **191**, 144–148.

149 *As the protons flow:* The rotation of ATP synthase is itself reducible to remarkable physical principles, in particular Brownian motion, in which the random movement of protons is used to drive its rotation in a form of ratchet motion. Many other biochemical processes tap into Brownian motion to achieve directional movement. See Oster G. (2002) Darwin's motors: Brownian ratchets. *Nature* **417**, 25. Like the bacterial flagellum, ATP synthase is another example of a circular wheel-like contraption in living things, albeit not for moving across surfaces, but for a rotating structure nonetheless.

149 *The changing shape of ATP synthase:* Phosphates are a chemical group with the formula PO_4^{2-}.

149 *The molecule so produced, ATP:* The phosphate bonds within ATP do not release energy when they are broken elsewhere in the cell (to break bonds requires energy). Instead, the small amount of energy needed to break the phosphate off ATP is more than made up for by the energy released when that phosphate binds to water after its release. The breakage of ATP is a *hydrolysis reaction*, and the net effect of all the bonds broken and made releases energy. A subtlety, but nevertheless an important one.

149 *If you think this is a trifling process:* The numbers are an estimate, as they depend on many factors. But roughly, you need about 2000

kilocalories of energy each day, that is, about three moles of glucose or about 1.8×10^{24} molecules of glucose. For each molecule of glucose shunted through the electron transport chain and onward, 36 molecules of ATP can be produced, so that's about 6.5×10^{25} molecules of ATP produced per day, or about 2.7×10^{24} per hour. Regardless of academic quibbling about the various conversions and efficiencies, the number is huge.

151   *Or is the architecture:* The conditions that gave rise to early proton gradients may have been hydrothermal vents. The early evolution of the process is discussed in Martin WF. (2012) Hydrogen, metals, bifurcating electrons, and proton gradients: The early evolution of biological energy conservation. *FEBS Letters* **586**, 485–493.

151   *We already know:* Imkamp F, Müller V. (2002) Chemiosmotic energy conservation with Na(+) as the coupling ion during hydrogen-dependent caffeate reduction by *Acetobacterium woodii. Journal of Bacteriology* **184**, 1947–1951.

151   *Nevertheless, protons are:* An excellent book exploring these early cellular processes and how the first gradients for energy acquisition might have formed is Lane N. (2016) *The Vital Question.* Profile Books, London.

152   *No group of people:* Boston PJ, Ivanov MV, McKay CP. (1992) On the possibility of chemosynthetic ecosystems in subsurface habitats on Mars. *Icarus* **95**, 300–308.

152   *Hydrogen gas can be ancient:* The link between serpentinization and life is discussed in Okland I et al. (2012) Low temperature alteration of serpentinized ultramafic rock and implications for microbial life. *Chemical Geology* **318**, 75–87.

153   *Microbial communities:* Spear JR, Walker JJ, McCollom TM, Pace NR. (2005) Hydrogen and bioenergetics in the Yellowstone geothermal ecosystem. *Proceedings of the National Academy of Sciences* **102**, 2555–2560.

153   *The core molecules:* Some of the proteins involved in the electron transfer chain are clearly ancient. For an early review, see Bruschi M, Guerlesquin F. (1988) Structure, function and evolution of bacterial ferredoxins. *FEMS Microbiology Reviews* **4**, 155–175. More recent studies have investigated their function and antiquity in deep-branching microorganisms. See, for example, Iwasaki T. (2010) Iron-Sulfur World in aerobic and hyperthermoacidophilic Archaea *Sulfolobus. Archaea,* 842639. The notion of an "iron-sulfur world" in

which these combinations of iron and sulfur atoms, perhaps in hydrothermal vent minerals, would provide the prebiotic conditions for the emergence of biochemistry and the electron transport process has been put forth by Günter Wächtershäuser, a particularly enthusiastic proponent of this version of early events. See, for example, Wächtershäuser G. (1990) The case for the chemoautotrophic origin of life in an iron-sulfur world. *Origins of Life and Evolution of Biospheres* **20**, 173–176.

154 *Place an electrode:* This has been a growing area of investigation. See, for example, Rowe AR et al. (2015) Marine sediments microbes capable of electrode oxidation as a surrogate for lithotrophic insoluble substrate metabolism. *Frontiers in Microbiology*, doi.org/10.3389 /fmicb.2014.00784; and Summers ZM, Gralnick JA, Bond DR. (2013) Cultivation of an obligate Fe(II)-oxidizing lithoautotrophic bacterium using electrodes. *MBio* **4**, e00420–e00412. doi: 10.1128 /mBio.00420-12.

154 *Elemental sulfur, thiosulfates:* The role and sheer scale of biogeochemical cycles is nicely explored in Falkowski PG. (2015) *Life's Engines: How Microbes Made Earth Habitable.* Princeton University Press, Princeton, NJ. For biogeochemical cycles in the marine environment, see Cotner JB, Biddanda BA. (2002) Small players, large role: Microbial influence on biogeochemical processes in pelagic aquatic ecosystems. *Ecosystems* **5**, 105–121.

154 *In a now seminal paper:* A whole book might be written about Broda. He was a communist sympathizer suspected to have been a KGB spy, code-named Eric, who may have been involved in passing information to the Soviets about British and American nuclear research. Anything to do with energy seems to attract interesting characters. Broda E. (1977) Two kinds of lithotrophs missing in nature. *Zeitschrift für allgemeine Mikrobiologie* **17**, 491–493.

155 *This anaerobic ammonia oxidation:* Strous M et al. (1999) Missing lithotroph identified as new planctomycete. *Nature* **400**, 446–449.

155 *The microbes, by using uranium:* The uranium becomes more "reduced," that is, it gains electrons as the electron acceptor. Lovley DR, Phillips EJP, Gorby YA, Landa ER. (1991) Microbial reduction of uranium. *Nature* **350**, 413–416.

155 *Combining sandwiches:* These reactions can be used to predict the amount of energy available, allowing scientists to then go into the environment to search for the potential microbes that might make use of

these energy-yielding chemicals. A good example is Rogers KL, Amend JP, Gurrieri S. (2007) Temporal changes in fluid chemistry and energy profiles in the Vulcano Island Hydrothermal System. *Astrobiology* 7, 905–932, which elegantly illustrates how life in extreme environments, potentially limited by energy, can be understood and predicted using the basic physics of Gibbs free energy in any given chemical reaction. Here we see how physics and the basic principles it elucidates can be used to enhance the predictive power of biological sciences.

156   *In anaerobic habitats:* Clearly where there is no energy, there can be no active life, but life also needs a basic level of energy to survive, and so for many organisms, even a little energy may be too little. The role of energy in limiting life is explored in Hoehler TM. (2004) Biological energy requirements as quantitative boundary conditions for life in the subsurface. *Geobiology* 2, 205–215; and Hoehler TM, Jørgensen BB. (2013) Microbial life under extreme energy limitation. *Nature Reviews Microbiology* 11, 83–94.

156   *However, why did the concentrations of oxygen:* Catling DC, Claire MW. (2005) How Earth's atmosphere evolved to an oxic state. *Earth and Planetary Science Letters* 237, 1–20.

158   *Giant tubeworms:* Two papers that explore this fascinating symbiosis are Cavanaugh, CM, Gardiner SL, Jones ML, Jannasch HW, Waterbury JB. (1981) Prokaryotic cells in the hydrothermal vent tube worm *Riftia pachyptila* Jones: Possible chemoautotrophic symbionts. *Science* 213, 340–342; and Minic Z, Hervé G. (2004) Biochemical and enzymological aspects of the symbiosis between the deep-sea tubeworm *Riftia pachyptila* and its bacterial endosymbiont. *European Journal of Biochemistry* 271, 3093–3102.

162   *The radiation produced:* Lin L-H et al. (2005) The yield and isotopic composition of radiolytic H_2, a potential energy source for the deep subsurface biosphere. *Geochimica et Cosmochimica Acta* 69, 893–903.

162   *Fungi that contained the pigment:* Dadachova E et al. (2007) Ionizing radiation changes the electronic properties of melanin and enhances the growth of melanized fungi. *PLoS ONE* 2, e457.

163   *In an elegant stroll:* Schulze-Makuch D, Irwin LN. (2008) *Life in the Universe: Expectations and Constraints.* Springer, Heidelberg.

163   *Kinetic energy:* By "some protozoa," I mean the ciliates, such as *Paramecium* species.

163   *Perhaps thermal energy:* Here I mean the direct use of thermal gradients. Photosynthesis using geothermally produced light (wavelengths

greater than approximately 700 nanometers) has been reported in hydrothermal vents, linking thermal environments to energy acquisition. However, such organisms use conventional photosynthetic apparatus that just happens to be using nonsolar photons (Beatty JT et al. [2005] An obligately photosynthetic bacterial anaerobe from a deep-sea hydrothermal vent. *Proceedings of the National Academy of Sciences* **102**, 9306–9310).

CHAPTER 9

167 *Samuel Taylor Coleridge:* Samuel Taylor Coleridge. (1834) *The Rime of the Ancient Mariner.*

167 *There is a lot of water on Earth:* Taken from the US Geological Survey website in December 2017.

168 *The quixotic intelligent interstellar cloud:* Astrophysicist Fred Hoyle, in an intriguing science-fiction story (*The Black Cloud,* published by William Heinemann in 1957), describes a giant sentient cloud that enters the Solar System and accidentally blocks sunlight from reaching Earth. The sentient being expresses some surprise that there could be life on this ball of rock.

168 *We know of no single organism:* It may not be outside the capacities of synthetic biologists and chemists to make self-replicating molecules—cells, even—that will operate in other solvents. But like artificially altered genetic codes and the incorporation of novel amino acids into proteins, these laboratory fabrications may tell us very little about whether such entities would emerge under natural processes.

168 *One of the most notable:* The phase diagram of water is remarkably complex, with unusual forms of water ice occurring under high pressures and temperatures as the hydrogen-bonding networks change in their orientation. See, for example, Choukrouna M, Grasset O. (2007) Thermodynamic model for water and high-pressure ices up to 2.2 GPa and down to the metastable domain. *Journal of Chemical Physics* **127**, 124506.

168 *This property is strange:* A gigapascal is a unit of pressure (one billion pascals). On Earth at sea level, atmospheric pressure is equivalent to 101,325 pascals.

169 *It inhabits the undergrowth:* Storey KB, Storey JM. (1984) Biochemical adaption for freezing tolerance in the wood frog, *Rana sylvatica. Journal of Comparative Physiology B* **155**, 29–36.

170   *Hydrolysis reactions:* An old paper but one that nevertheless presents some of the reactions illustrating the reactive nature of water is Mabey W, Mill T. (1978) Critical review of hydrolysis of organic compounds in water under environmental conditions. *Journal of Physical and Chemical Reference Data* 7, 383–415.

170   *By binding to the outside of proteins:* An excellent review on the role of water in the cell is Ball P. (2007) Water as an active constituent in cell biology. *Chemical Reviews* **108**, 74–108. As the author recognizes, our comprehension of how water works is changing quickly. However, the remarkably versatile and subtle roles of water in biochemistry are no longer in doubt.

171   *This arrangement allows the genetic code:* Robinson CR, Sligar SG. (1993) Molecular recognition mediated by bound water: A mechanism for star activity of the restriction endonuclease EcoRI. *Journal of Molecular Biology* **234**, 302–306.

172   *The ability of some proteins:* Klibanov AM. (2001) Improving enzymes by using them in organic solvents. *Nature* **409**, 241–246.

173   *However, there the similarities:* Benner SA, Ricardo A, Carrigan MA. (2004) Is there a common chemical model for life in the universe? *Current Opinions in Chemical Biology* **8**, 672–689.

173   *For example, it can dissolve:* The properties of ammonia have been known for a long time. See, for example, Kraus CA. (1907) Solutions of metals in non-metallic solvents; I. General properties of solutions of metals in liquid ammonia. *Journal of the American Chemical Society* **29**, 1557–1571.

174   *We leave the oceans:* A good discussion of some of these possibilities can be found in Schulze-Makuch D, Irwin LN. (2008) *Life in the Universe: Expectations and Constraints.* Springer, Berlin, which reviews some of the advantages and disadvantages of different solvents, but the authors conclude that no known solvent would be better than water, apart from, potentially, ammonia at low temperatures.

174   *The optimistic temperature:* For a suggestion of blimp-like creatures in the clouds of Venus, see Morowitz H, Sagan C. (1967) Life in the clouds of Venus. *Nature* **215**, 1259–1260. For sulfate-reducing bacteria that eat sulfate compounds in the Venusian atmosphere, see Cockell CS. (1999) Life on Venus. *Planetary and Space Science* **47**, 1487–1501. Sulfur also features in Schulze-Makuch D et al., Grinspoon DH, Abbas O, Irwin LN, Mark A, Bullock MA. (2004) A sulfur-based survival strategy for putative phototrophic life in the Venusian atmosphere.

Astrobiology **4**, 11–18. These thoughts are fun, and the reader should not take the authors of these papers to be expressing a genuine committed belief that Venus has life. Like many of these discussions, however, they can provide a backdrop to ask stimulating questions about our own biosphere. For example, here are just two questions that emerge from contemplating life on Venus: Can you have a persistent aerial biosphere on a planet when the surface is uninhabitable? Why don't we observe blimp-like balloon organisms floating in Earth's atmosphere?

174   *In an intriguing thought experiment:* Benner SA, Ricardo A, Carrigan MA. (2004) Is there a common chemical model for life in the universe? *Current Opinions in Chemistry and Biology* **8**, 672–689.

176   *And yet even here, they must get enough energy:* This is a calculation made for Mars, but the order-of-magnitude estimate is applicable to Earth (Pavlov AA, Blinov AV, Konstantinov AN. [2002] Sterilization of Martian surface by cosmic radiation. *Planetary and Space Science* **50**, 669–673).

177   *Here, even a radiation-resistant:* Dartnell LR, Desorgher L, Ward JM, Coates AJ. (2007) Modelling the surface and subsurface Martian radiation environment: Implications for astrobiology. *Geophysical Research Letters* **34**, I.02207.

177   *The formation of reactive oxygen species:* Price PB, Sowers T. (2004) Temperature dependence of metabolic rates for microbial growth, maintenance, and survival. *Proceedings of the National Academy of Sciences* **101**, 4631–4636; Lindahl T, Nyberg B. (1972) Rate of depurination of native deoxyribonucleic acid. *Biochemistry* **11**, 3610–3618; Brinton KLF, Tsapin AI, Gilichinsky D, McDonald GD. (2002) Aspartic acid racemization and age-depth relationships for organic carbon in Siberian permafrost. *Astrobiology* **2**, 77–82.

177   *Indeed, for chemical reactions:* Chemical disequilibria made from geologically active processes.

178   *Rivers of methane:* Lorenz R. (2008) The changing face of Titan. *Physics Today* **61**, 34–39.

180   *Using this chemical compound:* Stevenson J, Lunine J, Clancy P. (2015) Membrane alternatives in worlds without oxygen: Creation of an azotosome. *Science Advances* **1**, e1400067.

180   *They proposed that by reacting hydrocarbons:* McKay CP, Smith HD. (2005) Possibilities for methanogenic life in liquid methane on the surface of Titan. *Icarus* **178**, 274–276.

180 *These ideas have even received:* Strobel DF. (2010). Molecular hydrogen in Titan's atmosphere: Implications of the measured tropospheric and thermospheric mole fractions. *Icarus* **208**, 878–886.

180 *However, the presence of possible energy sources:* I say "most" because impacts on Titan's surface might generate local hydrothermal systems that warm the surface. Furthermore, a subsurface ocean on Titan might provide opportunities for prebiotic and biological processes.

181 *Then there are the ice caps:* The Kuiper Belt is a disc of objects beyond the orbit of Neptune. Although it is similar to the asteroid belt that lies between Mars and Jupiter, it is about twenty to two hundred times as massive.

182 *Look at the reaction scheme below:* See, for example, Klare G. (1988) *Reviews in Modern Astronomy 1: Cosmic Chemistry.* Springer, Heidelberg.

CHAPTER 10

185 *I absolutely deny being a Trekkie:* A bona fide *Star Trek* fan.

185 *creature is tracked down:* Tracking down an alternate life form in itself is an interesting problem that vexes astrobiologists: how do we detect life elsewhere with the minimum number of assumptions about its chemical composition? Of course, in *Star Trek*, the crew members merely change the settings on their tricorders, devices for scanning the world around them, to detect silicon-based life, but it is not clear how one would detect a silicon-based life form residing in rocks that average around 40 to 70 percent silicon.

186 *So hydrogen, with one proton:* Oganesson is named after Russian nuclear physicist Yuri Oganessian, who played a leading role in the discovery of the heaviest elements in the periodic table.

187 *This principle, that electrons:* Fermions are a group of subatomic particles, including the protons, which also exhibit this behavior. For the Pauli exclusion principle, see Massimi, M. (2012) *Pauli's Exclusion Principle: The Origin and Validation of a Scientific Principle.* Cambridge University Press, Cambridge. This book is a good place to look at this principle in more detail.

187 *Like the twins:* More exactly, no two fermions can have identical quantum numbers, the four numbers that define its state—the principal quantum number, the angular momentum in its orbital (known as the angular momentum quantum number), the availability of orbitals

(magnetic quantum number), and the spin quantum number. For particles that have a half-integer spin (such as electrons), the wave function that described its wavelike property is antisymmetrical, which means that if they are in an identical place, the two waves cancel each other out and the particles cease to exist, which is impossible. Therefore, either their spin or one of their other properties has to be different to prevent this occurrence.

188 *The remaining two:* Actually, these two electrons are split up, one each in two suborbitals, $2p_x$ and $2p_y$. As p_x and p_y exist at the same level and have the same energy, the electrons, which would really rather be separate, tend to occupy these different suborbitals.

188 *Two are placed in the next orbital up:* Like carbon, the two electrons in the outermost 3p orbital are in separate $3p_x$ and $3p_y$ orbitals.

189 *Resulting from all this versatility:* McGraw-Hill. (1997) *Encyclopedia of Science and Technology.* McGraw, New York.

190 *The silicon-silicon bond:* Alcock NW. (1990) *Bonding and Structure: Structural Principles in Inorganic and Organic Chemistry.* Ellis Horwood Ltd., New York. This information is also available from other standard chemistry texts.

190 *Silane (SiH_4):* Emeléus HJ and Stewart K. (1936) The oxidation of the silicon hydrides. *Journal of the Chemical Society* 677–684.

191 *These rocky silicates:* They include the layered sheets (the phyllosilicates), strings of compounds (the inosilicates), and individual silicate tetrahedra (the nesosilicates). A very good book on the wide number of silicates is Deer WA, Howie RA, Zussman J. (1992) *An Introduction to the Rock-Forming Minerals.* Prentice-Hall, New York.

191 *These photosynthesizing microbes:* Brzezinski MA. (1985) The Si:C:N ratio of marine diatoms: Interspecific variability and the effect of some environmental variables. *Journal of Phycology* **21**, 347–357.

191 *Plants also gather:* See, for example, Currie HA, Perry CC. (2007) Silica in plants: Biological, biochemical and chemical studies. *Annals of Botany* **100**, 1383–1389.

191 *Silica structures:* Müller WE et al. (2011) The unique invention of the siliceous sponges: Their enzymatically made bio-silica skeleton. *Progress in Molecular and Subcellular Biology* **52**, 251–281.

191 *The silicon and carbon compound:* Shiryaev AA, Griffin WL, Stoyanov E, Kagi H. (2008) Natural silicon carbide from different

geological settings: Polytypes, trace elements, inclusions. *9th International Kimberlite Conference Extended Abstract No. 9IKC-A-00075.*

192   *The atom seems to form:* Röshe L, John P, Reitmeier R. (2003) *Organic Silicon Compounds. Ullmann's Encyclopedia of Industrial Chemistry.* Wiley-VCH, Weinheim.

192   *Perhaps a black-and-white view:* Cells can be coaxed into incorporating silicon into organic bonds (Kan SBJ, Lewis RD, Chen K, Arnold FH. [2016] Directed evolution of cytochrome c for carbon–silicon bond formation: Bringing silicon to life. *Science* **354**, 1048–1051). However, engineering these capacities into life does not necessarily imply that if the tape of evolution were rerun, it would use these pathways. Artificial pathways successfully incorporated into cells are not necessarily those that would naturally be found and eventually used by life when it is faced with the selection pressures of a real planetary environment.

192   *Germanium is the next element:* Johnson OH. (1952) Germanium and its inorganic compounds. *Chemical Reviews* **51**, 431–469. This is admittedly an old paper, but a more modern knowledge does little to change the basic conclusion that germanium life forms seem unlikely.

193   *An imaginative, if highly unfamiliar suggestion:* Bains W. (2004) Many chemistries could be used to build living systems. *Astrobiology* **4**, 137–167.

193   *The liquid nitrogen would offer:* The silanes are chemical compounds consisting of one or several silicon atoms linked to each other or to one or several atoms of other chemical elements. They comprise a series of inorganic compounds with the general formula Si_nH_{2n+2}. They are similar to alkanes in carbon chemistry.

196   *Nevertheless, within these clouds:* Snow TP, McCall BJ. (2006) Diffuse atomic and molecular clouds. *Annual Review of Astronomy and Astrophysics* **44**, 367–414.

196   *Protons, electrons, gamma rays:* Ions are atoms that have gained or lost electrons and therefore have a negative or positive charge, respectively.

197   *Diffuse interstellar bands:* Herbig GH. (1995) The diffuse interstellar bands. *Annual Review of Astronomy and Astrophysics* **33**, 19–73.

197   *Carbon, produced by fusion:* See, for example, Kaiser RI. (2002) Experimental investigation on the formation of carbon-bearing molecules in the interstellar medium via neutral-neutral reactions. *Chemical Reviews* **102**, 1309–1358; Marty B, Alexander C, Raymond SN. (2013)

Primordial origins of Earth's carbon. *Reviews in Mineralogy and Geochemistry* **75**, 149–181; and McBride EJ, Millar TJ, Kohanoff JJ. (2013) Organic synthesis in the interstellar medium by low-energy carbon irradiation. *Journal of Physical Chemistry* **117**, 9666–9672.

198 *The six-atom rings of carbon:* There are a variety of discussions on polycyclic aromatic hydrocarbons and other complex carbon compounds. See, for example, Tielens AGGM. (2008) Interstellar polycyclic aromatic hydrocarbon molecules. *Annual Reviews in Astronomy and Astrophysics* **46**, 289–337; Bettens RPA, Herbst E. (1997) The formation of large hydrocarbons and carbon clusters in dense interstellar clouds. *Astrophysical Journal* **478**, 585–593; and Bohme DK. (1992) PAH and fullerene ions and ion/molecule reactions in interstellar circumstellar chemistry. *Chemical Reviews* **92**, 1487–1508.

199 *They form tubes:* Iglesias-Groth S. (2004) Fullerenes and buckyonions in the interstellar medium. *Astrophysical Journal* **608**, L37–L40.

199 *Astrochemists think:* Herbst E, Chang Q, Cuppen HM. (2005) Chemistry on interstellar grains. *Journal of Physics: Conference Series* **6**, 18–35.

200 *Just 390 to 490 light-years away:* IRC+10216 (CW Leo).

200 *Around the photosphere of the star:* Groesbeck TD, Phillips TG, Blake GA. (1994) The molecular emission-line spectrum of IRC +10216 between 330 and 358 GHz. *Astrophysical Journal Supplemental Series* **94**, 147–162.

201 *This molecule can take part:* Coutens A et al. (2015) Detection of glycolaldehyde toward the solar-type protostar NGC 1333 IRAS2A. *Astronomy and Astrophysics* **576**, article A5.

201 *Isopropyl cyanide:* Belloche A, Garrod RT, Müller HSP, Menten KM. (2014) Detection of a branched alkyl molecule in the interstellar medium: *iso*-propyl cyanide. *Science* **345**, 1584–1586.

201 *Equally extraordinary:* Pizzarello S. (2007) The chemistry that preceded life's origins: A study guide from meteorites. *Chemistry and Biodiversity* **4**, 680–693.

201 *Although the concentrations:* Sephton MA. (2002) Organic compounds in carbonaceous meteorites. *Natural Product Reports* **19**, 292–311; and Pizzarello S, Cronin JR. (2000) Non-racemic amino acids in the Murray and Murchison meteorites. *Geochimica et Cosmochimica Acta* **64**, 329–338.

201 *Why the discrepancy?* A discrepancy true for many other types of molecules.

202 *Sugars, the building blocks:* Deamer D. (2011) *First Life: Discovering the Connections Between Stars, Cells, and How Life Began.* University of California Press, Berkeley.

202 *Meteorites come from asteroids:* An astronomical unit is equivalent to the mean distance between the Sun and Earth.

202 *No mere blocks of ice:* Altwegg K. (2016) Prebiotic chemicals—amino acid and phosphorus—in the coma of comet 67P/Churyumov-Gerasimenko. *Science Advances* 2, e1600285,

203 *carbon chemistry:* The leap between simple carbon compounds and a self-replicating life form is immense, and we do not know if it is inevitable on any planet where there is liquid water and appropriate physical conditions. In this book, I do not address the question of how common life is in the universe. Rather, I am more concerned with whether, once it does emerge, it has universal characteristics. Whatever that spark or environmental condition was that allowed life to emerge, however likely or unlikely it was, it has no relevance to the observation that the conditions in which solar systems emerge produce a preponderance of organic compounds.

203 *The tendency of energetic environments:* For a description of this experiment, see Miller SL. (1953) A production of amino acids under possible primitive Earth conditions. *Science* 117, 528–529. A more recent study examining the results is Bada JL. (2013) New insights into prebiotic chemistry from Stanley Miller's spark discharge experiments. *Chemical Society Reviews* 42, 2186–2196.

204 *From above and below, young planets:* Chyba C, Sagan C. (1992) Endogenous production, exogenous delivery and impact-shock synthesis of organic molecules: An inventory for the origin of life. *Nature* 355, 125–132.

204 *Its methane lakes:* Raulin F, Owen T. (2002) Organic chemistry and exobiology on Titan. *Space Science Reviews* 104, 377–394.

204 *Its brown atmospheric haze:* Sagan C, Khare BN. (1979) Tholins: Organic chemistry of interstellar grains and gas. *Nature* 277, 102–107.

204 *Future robotic missions:* Lorenz RD et al. (2008) Titan's inventory of organic surface materials. *Geophysical Research Letters* 35, L02206.

205 *Of all the atoms available:* See Goldford JE, Hartman H, Smith TF, Segrè D. (2017). Remnants of an ancient metabolism without

phosphate. *Cell* **168**, 1–9, for a compelling hypothesis about a precursor to modern biology that may have worked without phosphorus.

207   *Familiar to most of us:* One landmark paper on this matter is Westheimer FH. (1987) Why nature chose phosphates. *Science* **235**, 1173–1178.

207   *The molecule ATP:* Maruyama K. (1991) The discovery of adenosine triphosphate and the establishment of its structure. *Journal of the History of Biology* **24**, 145–154.

207   *Strung down the backbone of DNA:* The classic paper of the elucidation of this structure is Watson JD, Crick FH. (1953) A structure for Deoxyribose Nucleic Acid. *Nature* **171**, 737–738, but of course much of the deeper understanding of the properties of DNA, including its phosphorus-containing backbone, has been developed since then and can be found in a vast literature.

208   *Two sulfur-containing amino acids:* Two molecules of the sulfur-containing amino acid cysteine. For the disulfide bridge, see Sevier CS and Kaiser CA. (2002) Formation and transfer of disulphide bonds in living cells. *Nature Reviews Molecular Cell Biology* **3**, 836–847.

208   *The carbon-fluorine bond:* Blanksby SJ, Ellison GB. (2003) Bond dissociation energies of organic molecules. *Accounts of Chemical Research* **36**, 255–263.

208   *In the tropics:* O'Hagan D, Harper DB. (1999) Fluorine-containing natural products. *Journal of Fluorine Chemistry* **100**, 127–133.

209   *It can be found in cells:* Baltz JM, Smith SS, Biggers JD, Lechene C. (1997) Intracellular ion concentrations and their maintenance by Na^+/K^+-ATPase in preimplantation mouse embryos. *Zygote* **5**, 1–9.

209   *In an article published:* Wolfe-Simon F et al. (2010) A bacterium that can grow by using arsenic instead of phosphorus. *Science* **332**, 1163–1166.

209   *The microbe:* Rosen BP, Ajees AA, McDermott TR. (2011) Life and death with arsenic. *BioEssays* **33**, 350–357.

210   *The estimated half-life:* A half-life is the time it takes for half of something, such as a chemical compound, to disintegrate.

210   *If you replace arsenate:* Fekry MI, Tipton PA, Gates KS. (2011) Kinetic consequences of replacing the internucleotide phosphorus atoms in DNA with arsenic. *ACS Chemical Biology* **6**, 127–130.

210 *Its uses are enigmatic:* Edmonds JS et al. (1977) Isolation, crystal structure and synthesis of arsenobetaine, the arsenical constituent of the western rock lobster *Panulirus longipes cygnus* George. *Tetrahedron Letters* **18**, 1543–1546.

211 *The energetic cost:* Reich JH and Hondal RJ. (2016) Why nature chose selenium. *ACS Chemical Biology* **11**, 821–841. This paper is a recapitulation of Westheimer's paper "Why Nature Chose Phosphates."

211 *It is an essential trace element:* See, for example, Blevins DG, Lukaszewski KM. (1998) Functions of boron in plant nutrition. *Annual Review of Plant Physiology and Plant Molecular Biology* **49**, 481–500; and Nielsen FH. (1997) Boron in human and animal nutrition. *Plant and Soil* **193**, 199–208.

212 *Elements like vanadium:* Many researchers are exploring the role of different elements in life, particularly the lesser-known elements. For vanadium and molybdenum, see, for example, Rehder D. (2015) The role of vanadium in biology. *Metallomics* **7**, 730–742; and Mendel RR, Bittner F. (2006) Cell biology of molybdenum. *Biochimica et Biophysica Acta* **1763**, 621–635.

212 *The imaginative may well raise their hands:* A popular take on this is in the 1969 novel by Michael Crichton, *The Andromeda Strain*. In the novel, a returning space capsule has been contaminated with a crystal-based life form that threatens to overrun the laboratory in which it has been contained and escape into the terrestrial environment. Eventually, and happily for the Earth, it mutates into a less malignant form of life.

213 *Mineral surfaces as places:* Mineral surfaces provide ordered structures for the assembly of polymers, which themselves become ordered and aligned. The possible role of minerals in the assembly of the first self-replicating genetic structures was discussed in Cairns-Smith AG, Hartman H. (1986) *Clay Minerals and the Origin of Life*. Cambridge University Press, Cambridge, and the area was reviewed nicely in Arrhenius GO. (2003) Crystals and life. *Helvetica Chimica Acta* **86**, 1569–1586.

CHAPTER 11

217 *Sometimes this is referred to:* This problem is nicely summarized by Mariscal C. (2015) Universal biology: Assessing universality from a single example. In *The Impact of Discovering Life Beyond Earth*, edited

by Dick SJ, 113–126; and Cleland CE. (2013) Is a general theory of life possible? Seeking the nature of life in the context of a single example. *Biological Theory* 7, 368–379.

217 *It is easy to get trapped:* I even feel uncomfortable with the term *physical principles*, despite using it prolifically in this book. What do we actually *mean* by *physical*? We just mean principles by which the universe works. The word *physical* tends to segregate physicists and other types of scientists, removing its neutrality and encouraging proudly defended disciplinary boundaries. Maybe we should just speak of *principles*. Nevertheless, I use the term because it does conveniently emphasize that we are talking about principles that pertain to matter and not other principles, like legal or moral ones.

218 *Although many people think:* It would be tempting to provide a proposed definitive list of things that are universal about life. However, I am reluctant because a person only has to make one false prediction, and the list becomes an example of the $N = 1$ problem, which is counterproductive. I find it more parsimonious to offer some broad suggestions. Defining such a list in greater detail and carrying out experiments to attempt to challenge it could produce worthwhile and interesting results and, over time, might generate a more robust list of characteristics at all scales of life—characteristics that most of us could agree were universal. See, for example, Cockell CS. (2016) The similarity of life across the Universe. *Molecular Biology of the Cell* 27, 1553–1555.

219 *The scaling laws:* West GB. (2017) *Scale: The Universal Laws of Life and Death in Organisms, Cities and Companies.* Weidenfeld & Nicolson, London.

220 *Perhaps, like DNA:* Benner SA, Ricardo A, Carrigan MA. (2004) Is there a common chemical model for life in the Universe? *Current Opinions in Chemistry and Biology* 8, 672–689.

221 *In our own Solar System:* See, for example, Grotzinger JP et al. (2014) A habitable fluvio-lacustrine environment at Yellowknife Bay, Gale Crater, Mars. *Science* 343, doi:10.1126/science.1242777.

222 *Substantial liquid water oceans:* There are many papers discussing the oceans of Europa, for example Hand KP, Carlson RW, Chyba CF. (2007) Energy, chemical disequilibrium, and geological constraints on Europa. *Astrobiology* 7, 1–18; Schmidt B, Blankenship D, Patterson W, Schenk P. (2011) Active formation of "chaos terrain" over shallow subsurface water on Europa. *Nature* 479, 502–505; Collins GC, Head JW,

Pappalardo RT, Spaun NA. (2000) Evaluation of models for the formation of chaotic terrain on Europa. *Journal of Geophysical Research* **105**,
1709–1716. For the moon Enceladus, see, for example, McKay CP et
al. (2008) The possible origin and persistence of life on Enceladus and
detection of biomarkers in plumes. *Astrobiology* **8**, 909–919; Waite JW
et al. (2009) Liquid water on Enceladus from observations of ammonia and ^{40}Ar in the plume. *Nature* **460**, 487–490; Waite JH et al. (2017)
Cassini finds molecular hydrogen in the Enceladus plume: Evidence
for hydrothermal processes. *Science* **356**, 155–159. And for the moon
Titan, see Raulin F, Owen T. (2002) Organic chemistry and exobiology
on Titan. *Space Science Reviews* **104**, 377–394.

222 *Even if they do, a confounding problem:* See, for example, Horneck G
et al. (2008) Microbial rock inhabitants survive hypervelocity impacts
on Mars-like host planets: First phase of lithopanspermia experimen-
tally tested. *Astrobiology* **8**, 17–44; and Fajardo-Cavazos P, Link L,
Melosh JH, Nicholson WL. (2005) *Bacillus subtilis* spores on artificial
meteorites survive hypervelocity atmospheric entry: Implications for
lithopanspermia. *Astrobiology* **5**, 726–736.

222 *It is premature:* We could detect these biospheres by looking for gases
such as oxygen in the planetary atmosphere. That in itself would tell
us something about the sorts of metabolisms that the alien life uses.
However, without a laboratory sample of this life, we will be limited in
the knowledge we can derive about its structure at the different levels
of its hierarchy that we have been discussing in this book.

223 *That discovery of this first so-called exoplanet:* The paper describing
this finding is Mayor M, Queloz D. (1995) A Jupiter-mass companion
to a solar-type star. *Nature* **378**, 355–359. The planet is named Pegasi
51b. Planets are generally named sequentially using letters.

224 *Many planets:* For one example of how these discoveries have reig-
nited new efforts to explain how the alignments of the planets in our
own Solar System came about, see Tsiganis K, Gomes R, Morbidelli
A, Levison HF. (2005) Origin of the orbital architecture of the giant
planets of the Solar System. *Nature* **435**, 459–461.

224 *Alongside the bounty of hot Jupiters:* Santos NC et al. (2004) A 14
Earth-masses exoplanet around μ Arae. *Astronomy and Astrophysics*
426, L19–L23.

224 *It has one-quarter the density:* Bakos GA et al. (2007) HAT-P-1b: A
large-radius, low-density exoplanet transiting one member of a stellar
binary. *Astrophysical Journal* **656**, 552–559.

224 *This inflated ball epitomizes:* Mandushev G et al. (2007) TrES-4: A transiting Hot Jupiter of very low density. *Astrophysical Journal Letters* **667**, L195–L198.

225 *Many of them are likely to be uninhabitable:* The first planets in the super-Earth-size range were found in 1992 orbiting the pulsar PSR B1257+12. Because a pulsar is the collapsed neutron star remnant of a supernova explosion, they are not thought to be habitable or to have oceans. Wolszczan A, Frail D. (1992) A planetary system around the millisecond pulsar PSR1257 + 12. *Nature* **355**, 145–147.

225 *Some are likely to be ocean worlds:* Charbonneau D et al. (2009) A super-Earth transiting a nearby low-mass star. *Nature* **462**, 891–894.

225 *The habitable zone:* The habitable zone is, like many concepts of its type, too simplified. One of Jupiter's moons, Europa, contains a giant ocean, and yet Jupiter is far outside the habitable zone. Europa's internal ocean is not maintained by heating from our Sun, but instead by the buckling and contortions caused by the moon's gravitational interactions with other Jovian moons. In this moon, there is liquid water far outside the habitable zone. Nonetheless, the habitable zone is useful because it allows us to identify a zone where we might find an Earth-like world around distant stars, a place with giant bodies of liquid water on its surface.

225 *It is slightly more aged:* We need not limit ourselves to the search for merely Earth-like planets. Some may be even more bizarre than our own home world. Just over twenty-two light-years away is a triple star system in which a red dwarf star is orbited by a double or binary star system made up of two K-type stars. Orbiting the red dwarf are at least two super-Earths in its habitable zone, Gliese 667Cb and c. If anything lives on these planets, it may be greeted regularly by the astonishing spectacle of a triple sunset. Even the writers of *Star Wars*, who so imaginatively conjured up a scene of Luke Skywalker on the moon Tatooine enjoying a double sunset, have been outdone by reality. Anglada-Escudé G et al. (2012) A planetary system around the nearby M Dwarf GJ 667C with at least one super-Earth in its habitable zone. *Astrophysical Journal Letters* **751**, L16.

225 *The sheer quantity of data:* Petigura EA, Howard AW, Marcy GW. (2013) Prevalence of Earth-size planets orbiting Sun-like stars. *Proceedings of the National Academy of Sciences* **110**, 19,273–19,278.

226 *And astronomers have brilliance:* Besides the methods I describe in the main text, there are other ingenious approaches. Gravitational lensing

uses the ability of massive objects in the universe to distort light to reveal the small blip in the light of a planet in orbit around a distant star, its light signature magnified briefly by the lensing caused by the focusing power of a massive object lying between it and the observers on Earth. Some exoplanets can be seen directly with telescopes. This is a little more challenging than the transit method, but by blocking out the light from the star, the little light reflected by a planet can be picked up and the pinpricks of individual planets discerned. We can achieve this remarkable feat using coronagraphs, telescopes with colossal sunshades to block out the glare of the central star and allow planets to be more easily detected. Even ground-based telescopes are successfully used to detect brown dwarfs, gassy planets about ten to eighty times the size of Jupiter. By observing them for long enough, we can even see changes in their atmospheres as gases swirl and heat under the influence of their star; in essence, astronomers have been able to observe weather on other planets. As you can imagine, though, direct detection works best with very large planets, thus explaining why the brown dwarfs are some of the more enticing candidates. There is now a legion of popular books describing the search for, and study of, exoplanets. Just one is Perryman M. (2014) *The Exoplanet Handbook*. Cambridge University Press, Cambridge.

229   *With all these bizarre new worlds:* "One is startled towards fantastic imaginings by such a suggestion: visions of silicon-aluminium organisms—why not silicon-aluminium men at once? wandering through an atmosphere of gaseous sulfur, let us say, by the shores of a sea of liquid iron some thousand degrees or so above the temperature of a blast furnace." Wells HG. (1894) Another basis for life. *Saturday Review*, 676.

230   *Some features may make other Earth-like worlds:* Heller R, Armstrong J. (2014) Superhabitable worlds. *Astrobiology* **14**, 50–66.

230   *Ecologists know well:* This is the so-called species-area relationship, a phenomenon itself amenable to modeling and physical interpretation. See, for example, Connor EF, McCoy ED (1979) The statistics and biology of the species-area relationship. *American Naturalist* **113**, 791–833.

233   *She fell to the ground:* Koepcke J. (2011) *When I Fell from the Sky*. Littletown Publishing, New York.

235   *There is no better place:* A good introductory book on this moon is Lorenz R, Mitton J. (2010) *Titan Unveiled: Saturn's Mysterious Moon Explored*. Princeton University Press, Princeton, NJ.

CHAPTER 12

239   *That the laws of physics and life:* There have been many gallant attempts to find "laws" in biology as distinct new insights. An example
is Bejan A, Zane JP. (2013) *Design in Nature: How the Constructal Law
Governs Evolution in Biology, Physics, Technology, and Social Organization.* Anchor Books, New York, which explores the idea that life
evolves toward solutions that enhance "flow" and proposes that this is
a unifying factor in all living systems. However, is this just a restate-
ment of the second law of thermodynamics? See also McShea DW,
Brandon RN. (2010) *Biology's First Law: The Tendency for Diversity
and Complexity to Increase in Evolutionary Systems.* University of Chicago Press, Chicago. Their "Zero-Force Evolutionary Law" proposes
that an increase in diversity and complexity in life observed over evolutionary time periods is a law. Is this merely a statement of the phenomenon that mutations and other changes in the code, DNA, will
inexorably produce diversity and variation without natural selection?
My own contention is that many attempts to find distinct biological
laws are indeed merely elaborated descriptions of phenomena in living things that have their origins in simpler physical principles and
might even be better formulated in these terms. Other efforts have
been made to use information theory and entropy to describe evolution. See, for example, Brooks DR, Wiley EO. (1988) *Evolution as
Entropy.* University of Chicago Press, Chicago. Examples such as this
provide potential mathematical and physical approaches to framing
evolutionary questions from the individual organism to the population scale.

239   *However, ultimately even the material:* The presence of life within
these inescapable laws of thermodynamics and the tendency toward
increased entropy is not a contradiction; nor is it a challenge to those
laws. See, for example, Kleidon A. (2010) Life, hierarchy, and the
thermodynamic machinery of planet Earth. *Physics of Life Reviews* **7**,
424–460.

240   *Consider this one from Jan Baptista van Helmont:* Quoted in Hall BK.
(2011) *Evolution: Principles and Processes.* Jones and Bartlett, Sudbury,
MA, 91. Also mentioned in a wider discussion on the origin of life is
Chen IA, de Vries MS. (2016) From underwear to non-equilibrium
thermodynamics: Physical chemistry informs the origin of life. *Physical Chemistry Chemical Physics* **18**, 20005.

240   *In the seventeenth century:* Gottdenker P. (1979) Francesco Redi and
the fly experiments. *Bulletin of the History of Medicine* **53**, 575–592.

241 *Just over sixty years after van Leeuwenhoek's discovery:* Needham JT. (1748) A summary of some late observations upon the generation, composition, and decomposition of animal and vegetable substances. *Philosophical Transactions of the Royal Society* **45**, 615–666.

241 *Transferring some of his gravy:* Before the days of health and safety, kitchen food, leftover gravy, and a smorgasbord of festering broths made excellent ways to advance the cause of science.

241 *After placing wetted seeds:* Wetted seeds provide nutrients for a whole range of microbes naturally attached to them to grow, so they were a favored way of getting small creatures growing in vials.

241 *He then showed:* Spallanzani L. (1799) *Tracts on the Nature of Animals and Vegetables.* William Creech et al., Edinburgh.

243 *Rather, at this infinitesimal scale:* This was a matter that also concerned Niels Bohr, who suggested that biology could not be readily reduced to physics, because, as is the case of quantum uncertainties, any observations of biology at the atomic level would disrupt an organism sufficiently (maybe even kill it), preventing us from taking reliable observations (Bohr N. [1933] Light and life. *Nature* **131**, 457–459). Bohr's thoughts have been somewhat overshadowed by the enormous number of methods developed since his time. With these methods, scientists can noninvasively study organisms without disrupting their functions sufficiently to make those observations questionable. Bohr made a related point that organisms are so complex compared with many physical systems that a reductionist approach to biology, particularly at the atomic level, is extremely difficult. For example, organisms' capability to take in gases and release waste products makes it problematic to define which atoms belong to an organism and which do not. However, even on this score, we might note the enormous strides made in biochemistry and biophysics since the 1930s to characterize life's processes at the atomic and even subatomic level. A recent discussion of Bohr's ideas in the light of new technology and knowledge can be found in Nussenzveig HM. (2015) Bohr's "Light and life" revisited. *Physica Scripta* **90**, 118001.

243 *We can write down simple equations:* For the less chemically inclined, this is a different mole from the furry ones I have already written about. In chemistry, a mole is the amount of a substance that contains an Avogadro's number of atoms, which happens to be 6.022×10^{23} (the number is obtained from the number of atoms in twelve grams of the isotope carbon-12). However, for those with a strange sense of humor, you can find websites that discuss how large a mole of moles would

be, and it is very large indeed. In fact, it is so large that such a mass of moles would be of interest to those who spend their time thinking about planet formation. There I will leave this point.

244 *In the machinery of the cell:* Smith TF, Morowitz HJ. (1982) Between history and physics. *Journal of Molecular Evolution* **18**, 265–282, is a well-written and thoroughly interesting paper that explores the interface between biology and physics, citing several authors and works that similarly see both congruence and differences between the two fields. The authors make a strong case for physical determinism at the level of biochemical pathways.

245 *Some bases (adenine and guanine):* These are "depurination" events because they cause the loss of purines (the bases adenine and guanine). They occur through hydrolysis reactions and are one of the main pathways that cause the disintegration of ancient preserved DNA. They also play a role in triggering cancer. Loss of pyrimidine bases (cytosine and thymine) can also occur, but the reaction rates are much slower.

246 *It was Per-Olov Löwdin:* If some mutations in DNA are caused by proton tunneling, namely, the consequences of quantum behavior, then we could perfectly well accept that some variations in organisms at the large scale have their origins in quantum-generated irregularities at the atomic scale. Proton tunneling in DNA base pairs, resulting in mutations, was discussed by Löwdin P-O. (1963) Proton tunnelling in DNA and its biological implications. *Reviews of Modern Physics* **35**, 724–732, and subsequently discussed by many others, for example Kryachko ES. (2002) The origin of spontaneous point mutations in DNA via Löwdin mechanism of proton tunneling in DNA base pairs: Cure with covalent base pairing. *Quantum Chemistry* **90**, 910–923.

246 *Now sometimes that proton:* These are tautomers, chemicals that have the same molecular formula and that readily interconvert.

247 *Nevertheless, I raise the question:* Lambert N et al. (2013) Quantum biology. *Nature Physics* **9**, 10–18; Arndt M, Juffmann T, Vedral V. (2009) Quantum physics meets biology. *HFSP Journal* **3**, 386–400; Davies PCW. (2004) Does quantum mechanics play a non-trivial role in life? *BioSystems* **78**, 69–79. The field of quantum biology may yet yield insights into how other effects at the quantum scale can influence biological processes at the larger scale. Photosynthesis is one process that may be influenced by quantum effects. For example Sarovar M, Ishizaki A, Fleming GR, Whaley KB. (2010). Quantum entanglement in photosynthetic light-harvesting complexes. *Nature Physics* **6**, 462–467.

248   *As Jacques Monod:* Written in the 1970s, when the first insights
into protein chemistry and the genetic code were being unraveled,
Monod's book is a beautifully written account of the behavior of life
at the molecular level and how this defines a difference with other
matter. However, even he succumbs to an astonished bewilderment
at how different life is to other matter: "On such a basis, but not on
that of a vague 'general theory of systems,' it becomes possible for us
to grasp how in a very real sense the organism effectively transcends
physical laws—even while obeying them—thus achieving at once the
pursuit and fulfilment of its own purpose" (Monod J. [1972] *Chance
and Necessity*. Collins, London, 81). If life obeys physical laws, it does
not transcend them at any level. Nevertheless, Monod's book explores
many of the general themes advanced by Smith and Morowitz (1982),
above, who say that the crucial difference between life and other
forms of matter occurs at the molecular level, specifically, the DNA
code that fixes errors and generates replicated variety.

248   *An atomic displacement:* However, defects, substitutions of atoms, and
other alterations can also sometimes be the source of new properties,
including greater strength.

248   *There is generally no way:* Crystals of substances that have a chiral
nature can replicate this chiral signature in subsequent crystals.
More-complex ideas for self-replicating crystals have been presented.
See, for example, Schulman R, Winfree E. (2005) Self-replication and
evolution of DNA crystals. In *ECAL 2005*, edited by M Capcarrere et
al. **LNAI 3630**, 734–743.

250   *Stephen Jay Gould:* For an entertaining and erudite insight into convergence and its possibilities, see Losos J. (2017) *Improbable Destinies:
How Predictable Is Evolution?* Allen Lane, London. Losos believes that
evolution is predictable, particularly among closely related lineages,
but that there is considerable scope for contingent events to shape
the course of evolution. My view is that contingency in the wonderful
diversity of life forms is constrained, but that this is not inconsistent
with the idea that physical solutions can be sufficiently varied to allow
for a mélange of living things.

250   *He recognized the underlying laws:* Gould SJ. (1989) *Wonderful Life:
The Burgess Shale and the Nature of History*. Hutchinson Radius, London, 289–290.

250   *He elaborated on this:* His book on the discovery of the Burgess Shale
explores his and his colleagues' exploits in revealing the hidden

treasures: Gould SJ. (1989) *Wonderful Life: The Burgess Shale and the Nature of History.* Hutchinson Radius, London.

251 *have a mouth:* This predictable structure was pointed out by Gould: "Bilaterally symmetrical creatures with heads and tails are almost always mobile. They concentrate sensory organs up front, and put their anuses behind, because they need to know where they are going and to move away from what they leave behind" (ibid., 156).

251 *Contingency is there:* A point of view expressed by Simon Conway-Morris in a counterpoint to Gould is Conway-Morris S. (1999). *The Crucible of Creation: The Burgess Shale and the Rise of Animals.* Oxford University Press, Oxford.

251 *Indeed, since Gould's paean to contingency:* A good summary of the current state of knowledge, which also situates it in modern genetic data, is given by Budd GE. (2013) At the origin of animals: The revolutionary Cambrian fossil record. *Current Genomics* **14**, 344–354.

251 *In some refinements:* For a summary of constraints imposed by the history of an organism, see Maynard Smith J et al. (1985) Developmental constraints and evolution. *Quarterly Review of Biology* **60**, 263–287. The authors also discuss how physical factors limit the spiral structures that can be used in shelled organisms, a particularly visual example of physical (biomechanical) factors driving evolution.

251 *The historical nuances:* But before the reader thinks I am capitulating, this statement is primarily directed at the fine details. For example, one could be overwhelmed by the vast diversity of skeletal structures in vertebrates, but even this diversity can be constrained into a few well-defined forms. A thorough discussion of this limitation can be found in Thomas RDK, Reif W. (1993) The skeleton space. A finite set of organic designs. *Evolution* **47**, 341–356.

253 *From that increase in surface area:* For a fascinating hypothesis on the factors that caused the transition from the Ediacaran to the Cambrian fauna, see Budd GE, Jensen S. (2017) The origin of the animals and a "Savannah" hypothesis for early bilaterian evolution. *Biological Reviews* **92**, 446–473, which provides a mechanism by which the transition out of the "flattened forms" of the Ediacaran could have occurred. A superb book on animal form and body plans is Raff RA. (1996) *The Shape of Life: Genes, Development, and the Evolution of Animal Form.* University of Chicago Press, Chicago. His book underscores that my statement about producing an invagination and turning a pancake into a more complex organism with internal organs is probably a little

flippant. The architecture and history of body plans and their phylogeny is a complex field still in dispute. However, my comment is simply designed to ask whether life really can be railroaded into a dead-end body plan from which it has no escape.

253   *Whether there are contingencies:* Even on present-day Earth, certain organisms, such as jellyfish, have a pancake-like architecture, where the cells of the body are close to the outside surface.

254   *the whale's indecision:* I highlight again here the difference between mutability in life's pathways and choices and the narrow limits of life curtailed by physical laws. The two are not contradictory. Life can have the flexibility to break free of past choices, but still be channeled into a limited set of forms.

254   *A growing compilation of evidence:* And these discoveries, particularly in evolutionary developmental biology, raise important questions about whether evolution is just a tinkerer that has no choice but to mess around with existing plans and formats or whether it can act more like an engineer, making something that is completely new and fashioned for its environment (Jacob F. [1977] Evolution and tinkering. *Science* **196**, 1161–1166). Clearly, evolution cannot start from scratch and must use what is there, but the restrictions in attempting new constructions may not be as great as once was thought. Jacob, in considering an alternative evolution, states unequivocally that "despite science fiction, Martians cannot look like us." However, the devil is in the details. What do we mean by "look like us"? If we mean exactly like us in detail, then we must agree with Jacob. If we mean using the same sorts of sensors, limbs to walk, and structures to support the organism against gravity, then they are likely to seem eerily like us (accepting, of course, that the term "Martians" is used metaphorically to mean aliens in general, not literal Martians, which if they exist at all today are likely to be microbial).

254   *But what about the emergence:* However, for a persuasive view that even multicellularity could emerge through the operation of simple physical principles (and may therefore be inevitable), see Newman SA, Forgacs G, Müller GB. (2006) Before programs: The physical origination of multicellular forms. *International Journal of Developmental Biology* **50**, 289–299.

255   *Although we cannot easily repeat evolution:* We can, of course, find vestigial organs and genetic signposts that might reveal contingency and past history at work in fashioning an organism. We might also compare past evolutionary pathways up to an extinction event with

those that occur afterward. For example, we might look at the evolution of the reptiles and compare them with the evolution of mammals after the end-Cretaceous. However, these are not controlled experiments in rerunning the sequence of evolution, since the environment has also changed, making it difficult, maybe impossible, to truly determine what is contingent and what is a consequence of altered conditions in which the organisms have been molded. We can more easily run evolutionary experiments in the laboratory and in certain well-controlled field settings to probe the role of contingency. For an excellent summary of research, from lizards to guppies and microbes, I recommend Losos J. (2017) *Improbable Destinies: How Predictable Is Evolution?* Allen Lane, London. However, experiments in the laboratory or even in the field still do not recapitulate the messy realities of Earth's history.

255   *We can deepen our ability:* I am reluctant to go as far as suggesting a periodic table of life, a suggestion made in McGhee G. (2011) *Convergent Evolution: Limited Forms Most Beautiful.* Massachusetts Institute of Technology, Cambridge, MA, because although life forms may be limited, the term *periodic table* gives the impression that the scope of the evolutionary process has a parity with the simplicity of the atomic structure of elements and a periodicity in structure akin to that of electron stacking. Although I discuss the physical principles at the heart of evolution, I do not claim that the result of the canalization of life by physical principles is a set of life forms as simple as atomic structure. Perhaps a better term would be something like *matrix of living forms.* Nevertheless, the idea of classifying life systematically in a tabular format broadly similar to the periodic table and according to some agreed-on parameters is an exciting one. Such a classification would be one way to formalize the limits in living form. Similar attempts may be valuable in categorizing niches. See Winemiller KO, Fitzgerald DB, Bower LM, Pianka ER. (2015) Functional traits, convergent evolution, and the periodic tables of niches. *Ecology Letters* **18**, 737–751.

256   *For instance, the fusiform, sleek body:* George McGhee states without ambiguity, "I predict with absolute confidence that if any large, fast-swimming organisms exist in the oceans of Europa—far away in orbit around Jupiter, swimming under the perpetual ice that covers their world—then they will have streamlined, fusiform bodies; that is, they will look very similar to a porpoise, an ichthyosaur, a swordfish, or a shark." Although large sea creatures in the oceans of Europa are less likely than microbes—if there is any life at all—his point about

the physical influence on convergent evolution at the level of the organism and its implications for a notion of universal biology is clear. See McGhee G. (2007) *The Geometry of Evolution.* Cambridge University Press, Cambridge, 148.

256   *This understanding might greatly simplify:* The observations of convergence at different levels of life's structural hierarchy also offer hope for simplifying rules of assembly of living things. For example, for a comparison with convergence at the level of whole organisms, see Zakon HH. (2002) Convergent evolution on the molecular level. *Brain, Behavior and Evolution* **59**, 250–261.

256   *In the finale to his seminal book:* Conway-Morris. S. (2004) *Life's Solution: Inevitable Humans in a Lonely Universe.* Cambridge University Press, Cambridge.

256   *Reductionism:* I am not a militant reductionist; nor is this book another tired attempt to reduce biology to its *simplest* physical principles. I echo Mayr's views that reductionism often destroys information at higher levels of a hierarchy, particularly in complex biological systems, since at higher levels, interactions between components often generate properties not manifest in their separate parts (see, for example, Mayr E. [2004] *What Makes Biology Unique?* Cambridge University Press, Cambridge, 67). Indeed, the investigation of self-organization and emergent complexity rests on the understanding that behavior at higher levels of biological hierarchy is not merely the sum of behaviors observed at lower hierarchies. As I have illustrated in Chapter 2 and elsewhere in this book, physical principles and equations can be applied to holistic biological entities such as flocks of birds or ant nests. A synthesis of physics and biology need not imply the age-old desire to break down biological phenomena into their tiniest parts, although historically this has often been the case and is often useful to do so.

INDEX

Charles Cockell is Professor of astrobiology at the University of Edinburgh and director of the UK Center for Astrobiology. He received his BSc in biochemistry and molecular biology from the University of Bristol and his doctorate in molecular biophysics from the University of Oxford. His interests encompass life in extreme environments, the habitability of extraterrestrial environments, and the exploration and settlement of space. He held a National Academy of Sciences associateship at NASA and then worked at the British Antarctic Survey and the Open University. He has published over three hundred scientific papers and numerous books, including a series on the conditions for liberty beyond Earth. His teaching has ranged from undergraduate education to Life Beyond, a program he established in prisons to engage prisoners in designs of space settlements. He received the Chancellor's Award for Teaching, the highest award for teaching at the University of Edinburgh. He is Chair of the Earth and Space Foundation, a nonprofit organization he established in 1994; the foundation supports expeditions that link space exploration with environmentalism.

Photograph courtesy of Johnny Watson.